INTRODUCING CHEMISTRY

The Salters' Approach

GRAHAM HILL JOHN HOLMAN
JOHN LAZONBY JOHN RAFFAN
DAVID WADDINGTON

To the teachers and students who helped us and took part in the trials of the Salters' Course
and to Francesca Garforth for her inspiration
and John Montgomery for his encouragement.

Heinemann Educational Books Ltd
Halley Court, Jordan Hill, Oxford OX2 8EJ

OXFORD LONDON EDINBURGH
MELBOURNE SYDNEY AUCKLAND
SINGAPORE MADRID ATHENS
IBADAN NAIROBI GABORONE HARARE
KINGSTON PORTSMOUTH NH

ISBN 0 435 64002 X

First published 1989
All rights reserved. Unauthorized duplication contravenes applicable laws.

Designed and typeset by The Pen and Ink Book Company Ltd,
Huntingdon, Cambridgeshire

Printed in Spain by Mateu Cromo.

CONTENTS

Introducing Chemistry: The Salters' Approach   iv

Metals

Introducing *Metals*	1
Looking at *Metals*	2
1 Poisonous metals	2
2 Using one metal to save another	4
3 Why are metals hard?	6
In brief *Metals*	7
Thinking about *Chemistry and metals*	8
1 What happens when iron rusts?	8
2 How do the reactivities of metals differ?	8
3 What is a chemical element?	9
4 How do chemists represent elements?	10
5 How can the properties of metals be modified?	10
Things to do	11

Drinks

Introducing *Drinks*	13
Looking at *Drinks*	14
1 Think before you drink	14
2 Tea – our national drink	16
In brief *Drinks*	17
Thinking about *Chemistry and drinks*	18
1 Where does our water come from?	18
2 How is drinking water purified?	18
3 How are alcoholic drinks made?	19
4 What happens in terms of particles when tea is made?	20
Things to do	21

Warmth

Introducing *Warmth*	23
Looking at *Warmth*	24
1 Heating our homes	24
2 Designing a gas burner	25
3 Fire!	27
4 Cleaning up the power stations	28
In brief *Warmth*	29
Thinking about *Chemistry and warmth*	30
1 What is energy?	30
2 What happens when fuels burn?	30
3 What happens when there isn't enough oxygen?	31
4 Why do burning fuels cause pollution?	32
Things to do	33

Clothing

Introducing *Clothing*	35
Looking at *Clothing*	36
1 Dyes for brighter clothing	36
2 Dry-cleaning	37
3 Protective clothes for firemen	38
4 Disposable nappies	39
In brief *Clothing*	41
Thinking about *Chemistry and clothing*	42
1 What kind of particles make good fibres?	42
2 How can you make stretchy fibres?	42
3 What are polymers?	42
4 What holds monomers together in a polymer chain?	43
5 Getting clothes clean	44
6 Wet clothes, dry clothes	44
Things to do	45

Food

Introducing *Food*	47
Looking at *Food*	48
1 Too much, too little, just right?	48
2 Slimming	49
3 Artificial sweeteners	50
4 Special diets for special people	50
In brief *Food*	52
Thinking about *Chemistry and food*	53
1 What's in food?	53
2 How do you keep warm?	55
3 How can you measure energy?	56
4 What foods give you energy?	56
5 How can you find out what is in food?	57
Things to do	58

Index   59

INTRODUCING CHEMISTRY:
THE SALTERS' APPROACH

About the approach

The Salters' approach is an entirely new way of teaching and learning chemistry for GCSE. It is an approach which goes further than simply putting chemistry into context.

- Each topic is based on aspects of everyday life.
- Chemical concepts and explanations arise naturally from the study of these everyday situations.

The approach was developed by a large and experienced group of school and university teachers. They felt that the Salters' approach needed an entirely new kind of textbook.

About this book

This book has been written to accompany the third year units of the Salters' Course. It is a valuable introduction to GCSEs in both Chemistry and Science. It contains five chapters, one for each unit, and each is divided into five key sections.

Introducing pages set the scene of the topic and raises some questions about what is in it. You should read this before starting the chapter.

The **Looking at** sections are short pieces on subjects relating to the theme of the chapter. You are not expected to read or learn about all of them. You can choose the ones you think are interesting. Or your teacher may ask you to read particular ones. These sections include questions to make you think more deeply about the main points.

In brief sections give you a straightforward summary of what you need to know to understand the topic

Thinking about pages explain the key chemical ideas that arise from each topic. When you read the *In brief* section, for example, you might realise that you need to look at a particular *Thinking about* section, or your teacher might ask you to study one. When you read this section for the first time you may decide to leave out the *Taking it further* boxes.

Things to do is a bank of activities for you to try and they include:
- *Activities to try*: investigations to do in the laboratory or at home
- *Things to find out*: questions to research from other books
- *Things to write about*
- *Points to discuss*: these are best discussed in small groups of three, four or five
- *Questions to answer*: these will help you when it comes to exams

In writing this book we set ourselves the task of showing you how chemistry affects your lives, and then helping you to understand the chemical ideas which explain these effects. In doing this, we had to learn a lot ourselves. We enjoyed writing the book; we hope you will enjoy using it.

Graham Hill, John Holman, John Lazonby, John Raffan, David Waddington

METALS

Figure 1 *The metal alloy which is used for the outer casing of this spaceship is strong, has a low density and is heat resistant.*

Introducing metals

You only have to look around you to see how important metals are in your life. There is a vast array of different metals which are used in different ways for different purposes.

People have not always had so many metals available to them. Most metals have to be extracted from rocks. You will find out more about how this is done later on in the course. Copper was the first metal to be used as it was the easiest to extract. Later people began to use bronze and iron. Aluminium, which is an important metal, was not produced commercially until 1886.

Now it is possible to obtain specially prepared mixtures of metals called **alloys** which have suitable properties for making everything from a spoon to a spaceship.

Explain what is wrong with the choices made by the people in figure 2.

Figure 2
Choosing the right metal for the right job is vital!

In this chapter you will see how

- metals have some similarities which enable us to know that an object is made from metal rather than any other material,
- metals also have many differences – in particular some metals corrode more easily than others,
- an understanding of the chemical reactions involved in corrosion can help to prevent it.

Looking at Metals

1 Poisonous metals

Figure 3 *Wealthy Greeks and Romans often used slaves to taste all their food and drink to see if it was safe to eat – a wise precaution!*

The ancient Greeks and Romans were in the habit of poisoning each other. Aristotle was executed by poisoning in 399 BC and it is thought that Agrippina poisoned Claudius with arsenic so that Nero could become emperor and then Nero himself poisoned Britannicus who was the son of Claudius. Clearly poisoning was a required political skill in those days. Because of stories such as these, there is a temptation to define a poison as a substance which kills someone but this is a very unscientific definition. Many substances which are essential in your diet can be poisonous if you take them in excess. Other more poisonous substances may be quite safe in very small quantities. So describing substances as either poisonous or safe is not very helpful. It is much more useful to know how poisonous substances are compared with each other.

Table 1 shows the comparative **toxicity** (how poisonous they are) of some substances.

The comparison is based on animal tests. The dose per kg of body weight which kills 50 per cent of the tested animals is called the Lethal Dose 50 per cent, or the LD50.

So the lower the LD50, the more lethal the substance. These values are probably useful for rough comparisons but we cannot be sure that the effect on humans is the same

Table 1

Substance	LD50/mg per kg	Toxicity
Alcohol	10 000	slightly toxic
Sodium chloride	4000	moderately toxic
DDT (insecticide)	100	very toxic
Nicotine	1	super toxic
Dioxin (impurity in a herbicide)	0.001	super toxic

as the effect on mice. Obviously we should even be concerned about doses which are many times smaller than the LD50.

In addition to the substances mentioned in table 1 there are a number of metals, sometimes called heavy metals, which are particularly **toxic**. Mercury, cadmium and lead are toxic heavy metals.

The poisonous nature of lead is probably due to its similarity to metals such as calcium which are essential to your health. For example, lead can replace the calcium in your bones and so stay in your body. So the concentration of lead in your body can continue to increase. For this reason lead is said to have a **cumulative effect**.

Lead poisoning can be treated with injections of a substance which will react more strongly with the lead than the compounds which are holding it in the body. This results in the lead being excreted from the body.

Figure 4 *Lead water pipes have been used since Roman times, and can be a cause of lead poisoning.*

Figure 5 *Busy urban roads like this can cause dangerous levels of lead to build up in the surrounding area.*

Causes of lead poisoning

Lead poisoning occurred in Roman times. If they drank slightly acidic wine from their lead goblets they would inevitably absorb some lead. The Romans also used **lead water pipes** (figure 4) and in fact their word for lead – *plumbum* has given us the word plumbing and also the symbol for lead, Pb.

Besides the continued use of lead pipes, the other major source of lead poisoning in the first half of this century was the production and use of **lead-based paints**. Lead is also used in the production of **car batteries**, so during the early days of their production the workers frequently developed lead poisoning. The use of lead paints is very much restricted and controlled now. Once the scientific evidence was collected about the harmful effects of lead, the working conditions in factories were made safe too.

Except for accidents, severe cases of lead poisoning are rare. However, there is concern about the toxic effects of small doses of lead over a period of time. There is evidence that young children may be particularly affected by relatively low doses of lead.

The major source of these small doses of lead is likely to be the **lead in petrol**. It is added to petrol in the form of a compound called **tetraethyl lead**. This is added to prevent the engine 'knocking' (which is caused by the mixture of air and petrol vapour exploding before the spark ignites it). The lead compound increases the efficiency of the engine but unfortunately the lead compounds pass out of the engine in the exhaust fumes (figure 5).

People living in urban areas are likely to be exposed to the greatest risk from these lead compounds. More and more countries are starting to use lead-free petrol (figure 6).

Figure 6 *Lead-free petrol is now becoming more readily available and many cars now use it.*

1. What sources of lead poisoning might exist in old houses?
2. Why is it particularly important that toys and cots should not be painted with lead-based paints?
3. When lead poisoning was first investigated in factories its presence was only recognised when black marks appeared in the gums of the workers.
Techniques are now available which can measure concentrations of lead in blood down to 0.00001g in 100g. Detecting lead at such low levels in the blood of people living in a city area may not be accepted as a sufficient reason for banning lead in petrol. What sort of arguments and evidence might be used for the case against banning lead in petrol?

2 Using one metal to save another

Iron is chosen to make so many objects because it is relatively cheap and it is strong (figure 7).

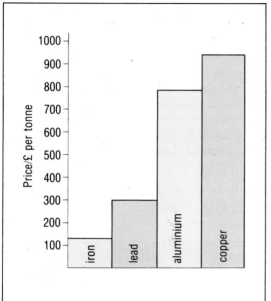

Figure 7 *Comparing the costs of some metals. Gold costs about £10 000 000 per tonne. Using the same scale as in the bar chart above, how high would the bar for gold be?*

Figure 8 *Tin cans are not just made of tin. They are made of iron which has been covered with a thin layer of tin. This makes them safe to put food in. Why?*

Figure 9 *A zinc dustbin is not just made of zinc. It is made of iron coated with zinc (called galvanised iron). Why is zinc used?*

But iron does have a serious disadvantage. It reacts very readily with air and water. This corrosion of iron is commonly called **rusting**. Other metals also **corrode** but the term rusting is used only for the corrosion of iron.

Other metals are often used to protect iron. The tin on the food can (figure 8) and the zinc on the dustbin (figure 9) protect the iron by keeping air and water away from it. Iron can only rust if it comes into contact with air *and* water.

The lumps of magnesium are obviously not covering the whole of the ship in figure 10 and so they must be protecting the iron in another way. Magnesium is more reactive than the iron. This means that the magnesium is corroded in preference to the iron. The magnesium is sacrificed to save the iron.

The magnesium lumps need to be replaced at regular intervals, but this is cheaper than replacing the ship!

Figure 10 *Lumps of magnesium are welded to the iron hulls of ships to protect them from rusting. You can see six magnesium bars in this picture. How do they protect the ship?*

Figure 11 compares the reactivity of different metals. This comparison is often called the **reactivity series**.

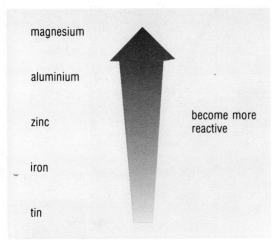

Figure 11 *Some metals in the reactivity series*

Aluminium is well above iron in the reactivity series and yet it does not appear to need special protection from corrosion. The secret is a very thin layer of aluminium oxide which forms on the surface of a freshly cut piece of aluminium when it is left in air (figure 12).

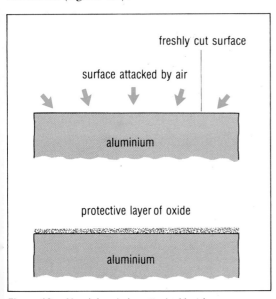

Figure 12 *Aluminium being attacked by air*

This layer of oxide sticks to the aluminium and prevents any more air attacking the rest of the aluminium. The aluminium has formed its own protective layer.

This layer can be artificially thickened using a process called **anodising**. You will find out more about this later.

1. Suggest reasons why strips of magnesium are used to protect the iron hulls of ships but not to protect iron gutters and drain pipes on a house?
2. The metal objects in figures 13 and 14 are protected from corrosion by thin layers of metal which also make them attractive. What metals do you think are being used and why is it a different metal in each case?

Figure 13

3. When metals corrode they usually change into the metal oxide. In the case of aluminium this oxide then protects the rest of the aluminium. Suggest a possible reason why iron oxide does not protect iron in the same way.

3 Why are metals hard?

It is possible to make solid objects from wood, plastic, pottery or metal. They are all **hard** in the sense that if you are hit by an object made from one of these materials it hurts. Obviously the one you would least like to be hit with is the metal object, but bumping into objects is not a very scientific way of comparing the materials they are made of!

But there are several ways of defining the meaning of hard.

One meaning is to do with the hardness of the surface – which material is most difficult to scratch or cut?

Another meaning is strength (figure 15) – if one end of the object is fixed and the other end is pulled, would it stretch or break?

A third meaning might be to do with what happens if you try to bend the material – does it bend or does it snap?

If you consider those metals such as aluminium, copper, iron and tin which are used to make objects, then in general these metals compared to wood, plastic or pottery:

♦ have surfaces which are more difficult to scratch or cut,
♦ are stronger when tested by pulling,
♦ are more bendy and are less likely to snap when bent.

But why are some metals harder than other materials? You will get a clue by looking closely at their surfaces (figure 16).

Each grain is a single **crystal** of the metal. When a metal is first extracted from its **ore** it is a liquid. As the liquid cools it starts to solidify as tiny crystals form in different parts of the liquid. As the metal continues to cool the crystals grow until they meet the other crystals around them and then the whole metal is a solid mass of tiny crystals.

The hardness of the metal is linked to its crystalline form and how strongly the crystals are held together.

But some metals are much harder than others and even different pieces of the same metal can have different hardnesses (figure 17).

Figure 15 The steel cables used by cranes like this one are strong enough to lift and hold very heavy objects.

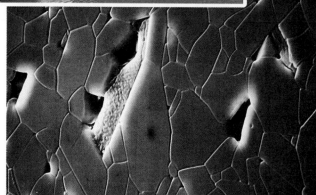

Figure 16 This photograph shows the magnified surface of aluminium. It shows that it is made up of tiny grains tightly packed together.

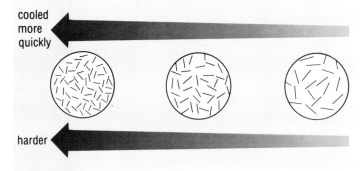

Figure 17 Cooling liquid metals

1 How do you think the hardness of the metal is influenced by the number of crystals and their size?
2 Draw a table which lists the three properties of metals mentioned in this section. Include additional columns for wood, plastic and pottery. Fill in next to each property how you think these other three substances differ from metals.

In brief
Metals

1. Metals are used in the construction of buildings, bicycles, cars, trains, aeroplanes and countless everyday objects used in the home, at work and in leisure pursuits. Metals are used when their properties suit the task and the cost of producing them matches what people are willing to pay.

2. Metals have particular physical properties which make them different from other materials such as plastic, wood or pottery.

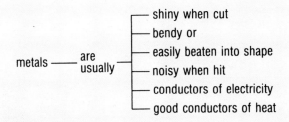

Figure 18 *Physical properties of metals*

Not all metals are good examples of all of these properties, but usually each one shows enough of them for you to know that it is a metal.

3. Although metals have a lot of similarities, they also have a lot of differences. These differences are important when deciding which metal to use for which purpose. For example, some metals are stronger than others, some are heavier than others and some are more resistant to attack by air.

4. When a metal is attacked by air, it has been **corroded**. Corrosion is a **chemical change** because the metal reacts with the oxygen in the air (and sometimes water) to form a new substance.

metal + oxygen → new substance
(and water)

In the case of iron, the new substance formed is iron oxide, called rust.

5. Some metals corrode more readily than others. They are more reactive. The order of reactivity of some of the metals is shown below. It is called the metal **reactivity series** (figure 19).

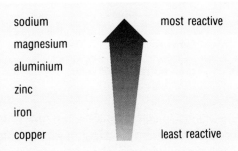

Figure 19 *Some metals in the reactivity series*

6. To prevent the corrosion of a metal you can:
   - keep air and moisture away from the metal,
   - change the properties of the metal by alloying it,
   - fix small pieces of a more reactive metal to its surface.

Ways of preventing corrosion by covering the metal

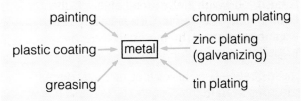

7. An **alloy** is a mixture of two or more metals (figure 20). It is made by mixing the metals together as liquids and then allowing them to cool and solidify.

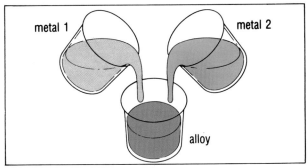

Figure 20 *Making an alloy*

Alloys have different properties from any of the metals they contain.

8. A pure metal is an element which means that it cannot be broken down into anything simpler. For example, pure copper is just copper and nothing else.

In the same way pure non-metals such as oxygen and carbon are elements because they cannot be broken down into anything simpler.

Each element can be represented by a **symbol**. The symbol is one or two letters from the name of the element. For example, magnesium is Mg, iron is Fe (which is based on the Latin name for iron *ferrum*) and oxygen is O. The symbols are a type of shorthand for the elements; they are understood everywhere in the world.

The product of corrosion, such as rust, is a **compound** because it contains more than one element (iron and oxygen).

Thinking about
Chemistry and metals

1 What happens when iron rusts?

If you investigated what happens when iron rusts you would find that:

- iron nails will only rust when both water and air are present,
- the iron seems to be 'eaten away' by the rusting, but if you weigh the iron that is left with the rust which has been formed, there has been an increase in mass,
- rust looks very different from the original iron.

These observations tell us that iron must combine with something from water and air to make a new substance which we call rust (figure 21).

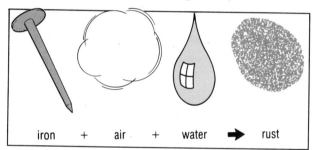

iron + air + water ➔ rust

Figure 21

2 How do the reactivities of metals differ?

Look at figures 22 – 24.

Figure 22 *Sodium is so reactive that it must be kept away from air and so it is stored under oil.*

Figure 23 *Iron which is not protected from a damp atmosphere will rust.*

Figure 24 *Gold objects such as this mask from an Egyptian tomb will stay shiny for ever.*

You can see from these photographs that sodium is more reactive than iron, whereas gold is much less reactive than either of the other two metals.

Similar differences are found if the reactions of metals with water are investigated (table 2).

Table 2

Sodium	reacts violently with water and gives off a gas
Calcium	reacts steadily with water and gives off a gas
Magnesium	reacts very slowly with water and gives off a gas
Iron	reacts with water and air after a day or so to form rust
Copper	does not react with water

If the gas which is given off is collected (this is most easily done with calcium) and tested with a lighted splint, it will pop or explode. This shows that it is hydrogen.

Taking it further

When sodium or calcium or magnesium react with water, the gas given off is hydrogen. The solution which is left turns litmus blue showing that it is an **alkali**. Alkalis are metal hydroxides, so the reaction of, for example, calcium with water can be summarised by the word equation:

calcium + water → calcium hydroxide + hydrogen

Similar equations can be used to summarise the reactions of sodium and magnesium with water.

Evidence such as how metals react with air and water helps us to build up the reactivity series. Figure 25 shows a more complete series.

Figure 25 *Building up the reactivity series*

sodium
calcium
magnesium
aluminium
iron
zinc
copper
gold

most reactive

Hydrogen and oxygen are both elements. Pure water (H_2O) is not an element it is a compound in which hydrogen and oxygen are joined together. It is possible, although not so easy, to get the hydrogen and oxygen out of water, but it is not possible to get anything out of hydrogen or to get anything out of oxygen.

You cannot tell by just looking at a substance that it is an element. It is just by experience that you learn which substances are elements and which are not.

Everything around you is made up of elements. You are, the chair you are sitting on is, the air you breathe, the food you eat – in fact the whole universe is made up of elements and combinations of elements.

However there is only a limited number of elements. The sun, the moon, the planets, Halley's comet; none of them can contain any elements which are not found on the earth (figures 28 and 29).

3 What is a chemical element?

A pure metal such as copper is an element – it contains nothing but copper. Brass is not an element because it contains a mixture of copper and zinc – it is an alloy of two elements.

▲ **Figure 26** *Copper is an element (figure 26) but brass is made from a mixture of copper and zinc (figure 27). Brass is an alloy.*

Figure 27 ▶

Figure 28

Figure 29 *Moon rock is made from elements which are all found on the earth too.*

4 How do chemists represent elements?

Chemical names can be very long. They also vary from one language to another. This could cause considerable confusion.

Figure 30 *What are in these bottles?*

So chemists use a system of shorthand for the elements. They write a symbol which is either one or two letters instead of writing out the whole name. For example,

O stands for oxygen
and Mg stands for magnesium.

These symbols are part of an international language. So a chemist, or chemistry student, anywhere in the world will know that Mg is magnesium. This is just like a pianist in any country being able to recognise the way Chopin wrote down his music or a guitarist in any country recognising the chord symbols used to represent a tune.

An element is represented by the first letter of its name, or if there are two or more elements starting with a particular letter, then another letter from its name is also used:

Carbon is represented	by C
Calcium	by Ca
Cobalt	by Co
Chlorine	by Cl
Chromium	by Cr

Sometimes the symbol is derived from the Latin name for the element:

Sodium (*Natrium*)	by Na
Copper (*Cuprum*)	by Cu
Iron (*Ferrum*)	by Fe

5 How can the properties of metals be modified?

Electricians or electronics engineers need to be able to join components to copper wires. The joint must conduct electricity and so it must be made of metal. When the metal is heated gently it should melt without melting the copper or damaging the components and when it cools it should form a solid metal joint.

Solder is an alloy of lead and tin and has the ideal properties for joining electrical components. It melts at a lower temperature than either of the pure metals and so doesn't damage the components (figure 31).

Figure 31 *Solder (an alloy of lead and tin) can be melted with a hot soldering iron and used to join electrical components without damaging them.*

Most metals in everyday use are in fact alloys. By adding another metal, or occasionally a non-metal, to the pure metal its properties can often be changed to make it suitable for different purposes. Figures 32 and 33 show some common alloys.

◄ **Figure 32** *This steel scalpel is made from an alloy of iron and carbon. It is stronger than pure iron.*

Figure 33 *This aeroplane is made of a light and strong titanium alloy.* ▼

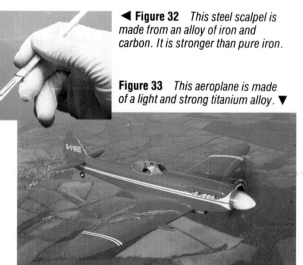

Things to do

Things to try out

1. Identify five objects in your home which you think are made of metal. Without damaging the objects carry out tests on them to decide if they are definitely made of metal. Draw a table with the names of the objects in one column. In a second column list the properties which have helped you to decide that they are made of a metal.

2. A food manufacturer produces tinned vegetables and tinned fruit. The vegetables are stored in salt solution and the fruit in sugar solution. Design and carry out your own investigation into the effect of these two solutions on dented tin cans.

Things to find out

3. With the help of older members of your family identify three objects which used to be made of metal and are now made of plastic. By discussing these objects with your family try to decide whether metal is no longer used for each of these objects because it is too expensive or because the plastic has more suitable properties.

4. Whenever a way of extracting a metal from its ore was discovered the metal became available for making useful objects. By consulting other books find out when some of the more important metals were first extracted and construct a time chart, possibly starting with copper. Metals which could be included are: aluminium, chromium, iron, lead, tin, tungsten and zinc.

Things to write about

5. Figure 34 shows the composition of the earth's crust. Copper is one of the metals in the 'other elements' section. It is thought that about 0.007 per cent of the earth's crust is copper.

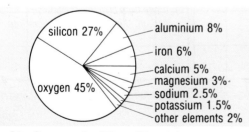

Figure 34 *Components of the earth's crust*

Most copper ores contain no more than 1 per cent of copper.

The consumption of copper metal in the USA works out at about 8 kg per head of population each year. In India it is about 0.1 kg per head of population.

The population of the USA is about 240 million and the population of India is about 770 million.

Write a magazine article, with an eye-catching title, which discusses what you think is important about these figures.

6. There have been particular incidents of mercury poisoning in the sea off Japan and in the Great Lakes of America. Using other books find out what you can about the source of the mercury poisoning and how it came to affect people.

Making decisions

Table 3

Metal (scale 1 – 3) (1 = most)	A	B	C	D
Strength	1	3	2	1
Flexibility	3	1	2	2
Corrodability	2	2	2	3
Density	medium	low	low	medium
Cost	medium	cheap	cheap	expensive

7. Which of the metals, A to D, would you choose to make:
   a) a pressurised can for deodorant spray,
   b) a fret for a guitar,
   c) a toothpaste tube,
   d) the metal part of a diamond stylus?

Points to discuss

8. The average consumption of milk in the UK is 240 pints per person per year. If you assume that 80 per cent of it is delivered in bottles with aluminium tops and that the mass of a bottle top is about 0.2 g, estimate the total mass of aluminium used to make milk bottle tops per year. What are your reactions to these data? What other data do you think you need to know before you can come to a definite view on the use, conservation and recycling of aluminium?

9. A motor car exhaust pipe which is made of stainless steel will corrode much more slowly than one made from normal steel. However, it costs about twice as much as a normal exhaust pipe. How would you decide whether or not to use a stainless steel exhaust pipe?

Questions to answer

10 A Aluminium
    B Copper
    C Steel
    D Chromium
    E Sodium

Choose from the metals A to E, the one which is
a) an alloy,
b) NOT an element,
c) used to make milk bottle tops,
d) used to protect iron,
e) a pink/brown colour,
f) the most reactive with water,
g) represented by the symbol Cu,
h) corroded to form rust.

11 Look at the Periodic Table below (figure 35). Each square contains the symbol for an element. The table can be thought of as an arrangement of all of the elements which results in similar elements being near each other.

a) Nickel (Ni) is used to make coins. Locate it in the table and find and name two other metals which are used to make coins.
b) Gold (Au) is used to make jewellery. Locate it in the table and find and name two other metals which are used for the same purpose.
c) Potassium (K) reacts vigorously with cold water. Locate it in the table. Find and name another metal which reacts in the same way with water.
d) Cobalt (Co) is magnetic. Locate it in the table. Find and name another metal which is also magnetic.

12 State, with reasons, whether each of the following substances is an element, an alloy or a compound:

brass, copper, lead, water, duralium, copper(II) oxide, rust, iron, solder, magnesium oxide, oxygen.

13 Based on their reactions with oxygen it is possible to arrange the following metals in the order of reactivity indicated. The most reactive metal is at the top of the list.

sodium
magnesium
iron
copper

Explain which of the following reactions you would predict to be likely to occur and which would not be likely to occur.

magnesium + copper(II) oxide → copper + magnesium oxide

iron + sodium oxide → sodium + iron oxide

group I	group II											group III	group VI	group V	group VI	group VII	group 0
					H												He
Li	Be											B	C	N	O	F	Ne
Na	Mg											Al	Si	P	S	Cl	Ar
K	Ca	Sc	Ti	V	Cr	Mn	Fe	Co	Ni	Cu	Zn	Ga	Ge	As	Se	Br	Kr
Rb	Sr	Y	Zr	Nb	Mo	Tc	Ru	Rh	Pd	Ag	Cd	In	Sn	Sb	Te	I	Xe
Cs	Ba	La	Hf	Ta	W	Re	Os	Ir	Pt	Au	Hg	Tl	Pb	Bi	Po	At	Rn
Fr	Ra	Ac															

Figure 35 *Periodic table*

DRINKS

Figure 1 *The water collected from this stream will be used for drinking and cooking. The stream may also be used for washing and sanitation so there is great danger that it will become infected.*

Introducing drinks

People go on hunger strikes to make a strong protest. One of the most famous hunger strikes was by the suffragettes 70 years ago. The suffragettes wanted votes for women. Later Mahatma Gandhi went on hunger strike against the British authorities in India. Gandhi was trying to win freedom for his country. Unlike hunger strikes, we never hear of thirst strikes.

Everyday you lose two litres of water through sweating and urinating. So everyday you need to replace this water. After breathing, drinking is the most necessary activity for sustaining life. People can live for 50 to 60 days without eating but will die after five to ten days without water.

For thousands of years, people took their drinking water from the nearest river or stream. As towns and cities got larger, this became very unhealthy. All kinds of domestic and industrial waste was being dumped in the same streams from which people took their drinking water. This led to outbreaks of disease. For example, 50 000 people died from cholera in England in 1831 when the Thames and other rivers became infected. You can now take it for granted that in Britain there will always be clean and safe water coming from the taps. However, this is not so in many other countries.

There are huge numbers of people in the world who do not have their own supply of water. Some people have to walk many miles each day to collect water. Others may only have a water supply which is polluted and unclean.

◀ **Figure 2** *By building protected wells, like this one, for drinking water, the risk of infection is greatly reduced.*

In this chapter you will see how

- ◆ chemistry helps us to clean and purify water,
- ◆ chemistry helps us to produce and manufacture different kinds of drinks.

Looking at Drinks

1 Think before you drink

What does the law say about alcohol?
- It is against the law to sell alcoholic drinks to anyone under the age of eighteen.
- No-one under eighteen can work in a bar.
- No-one under fourteen is allowed in a bar or an off-licence.

In spite of these laws, 91 per cent of seventeen year old boys and 35 per cent of seventeen year old girls admitted drinking illegally in pubs in a recent survey.

Breath test and breath test failures (1986)					
Car drivers	Involved in incident	Tested	Failed	Failed as % Involved	Tested
under 17	981	113	38	3.9%	33.6%
17 - 19	25 498	6892	1057	4.1%	15.3%
20 - 24	49 123	11 421	2724	5.5%	23.9%
25 - 28	31 668	6235	1590	5.0%	25.5%
29 - 34	38 375	10 622	1577	4.1%	23.8%
All ages	290 580	49 559	10 014	3.4%	20.2%

1 What percentage of all drivers involved in an incident failed a breath test?
2 Work out what percentage of 20 - 24 year olds involved in incidents were tested.
3 List two arguments for and two against random breath testing.

What happens if you drink alcohol?
Although most young people learn to drink moderately and safely as they get older, some teenagers drink too heavily. This causes serious problems at home and at school.

Young people seem to be affected by alcohol more easily than adults. Drinking begins to affect their judgement well before they reach the legal limit for drinking and driving (figure 3). At the legal limit, a young driver is five times more likely to have an accident than when he or she has not been drinking.

Be very careful about when and how much you drink. Look closely at what the Law says. Alcohol is a **strong drug**. It can affect your health and your whole life.

Figure 3 Testing a driver's breath to see if he has consumed too much alcohol to drive.

How much alcohol do drinks contain?

Different drinks contain different amounts of alcohol. Table 2 on page 20 shows the percentage of alcohol in different drinks. But remember the volume of the drink is important as well as the percentage of alcohol in it. Figure 4 opposite shows the normal measures sold in pubs. Each of these contain about the same amount of alcohol.

What happens if you drink alcohol?

After drinking, liquid passes quickly from the mouth to the stomach and then into the intestine (figure 5). Only small particles can pass through the lining of the stomach into the blood. Alcohol particles are small enough to do this. So you feel the effects of alcohol very soon after drinking it.

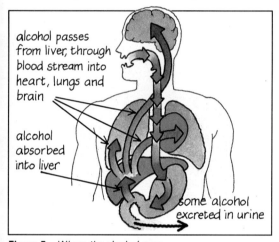

Figure 5 *Where the alcohol goes*

If you eat at the same time, alcohol is absorbed by the food. So it takes longer to get into your blood.

From the small intestine, the blood first passes to the liver. Here a small amount of alcohol is removed by the liver and excreted in the urine. As it passes through the liver, some alcohol reacts with oxygen in the blood. This produces carbon dioxide and water and provides your body with energy:

alcohol + oxygen → carbon dioxide + water + energy

Most of the alcohol, however, passes through the liver unchanged and gets to the heart, to the lungs (where small amounts pass into the breath) and to the brain. Having gone round the body, some of it returns to the liver and the cycle begins again.

Figure 4

How does alcohol affect your body?

In general, the more you weigh, the less you are affected by alcohol. This is because the alcohol can spread throughout a large volume of your body. It is less concentrated and therefore has a reduced effect.

Men and boys are also less affected by alcohol than women and girls of the same weight. This is because males have larger livers than females, so they can remove alcohol from their bodies more quickly.

In small amounts, drink can make some people chatty and funny. Others become aggressive.

Who do you think will be affected more by drinking alcohol, a stocky middle-aged man or a slim young woman? To answer this question, you have to consider four main factors:

- how much they have drunk,
- whether they have had a meal with their drink,
- their weight,
- their sex.

Alcohol affects your brain and your liver more than other parts of your body.

a) **Alcohol affects your brain**
   Alcohol is a depressant. This means that it depresses (slows down) your thinking. So, it slows down your judgement, your self-control and your skills. You will react more slowly to danger and operate machinery and cars with less care. In the UK, one in every three drivers killed in road accidents are over the legal limit. Excessive drinking can also cause brain damage and psychiatric (mental) problems like depression.

b) **Alcohol affects the liver**
   The liver is like a car with only one gear. It works best at one steady rate. Too much alcohol over a number of years can lead to hepatitis (inflammation of the liver) and cirrhosis (scarring of the liver).

---

4 Explain why the concentration of alcohol in someone's body is affected by:
- how much they have drunk,
- whether they have had a meal with their drink,
- their weight,
- their sex.

2 Tea – our national drink

Tea is Britain's national drink. On average we drink 3.7 cups of tea a day. Figure 6 below shows the popularity of different drinks and the average number of cups we drink per day.

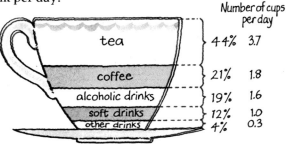

Figure 6 *The most popular drinks*

Table 1

Year	Cost of 1 kg of tea	Average weekly earnings
1700	£1.00 to £4.00	12.5p
1800	£1.80 to £5.50	25p
1900	33p	50p
Today	£3.50	£150

1. Suggest reasons why tea was much cheaper in 1900 than in 1800.
2. Would you think that more people can afford to buy tea today than in 1900?
3. At what date(s) was tea priced as a luxury? Give your reasons.

How is tea produced?

Figure 7

1 Plucking Tea leaves are plucked from small tea bushes (figure 7).

Figure 8

2 Withering Plucked leaves are spread on racks in a current of *warm* dry air and left to 'wither' (figure 8). This allows the moisture to evaporate from the leaves.

Figure 10

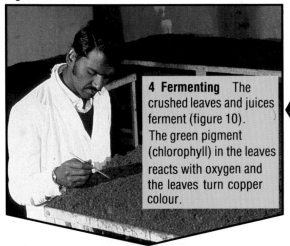

4 Fermenting The crushed leaves and juices ferment (figure 10). The green pigment (chlorophyll) in the leaves reacts with oxygen and the leaves turn copper colour.

Figure 9

3 Rolling The withered leaves are crushed by rolling, cutting and tearing (figure 9). This releases juices from the leaves.

5 Drying The fermented leaves are dried in *hot* air. The leaves turn black. The leaves are then **sorted**, **tasted** and **packed**.

1. Why is *warm* air used in the 'withering' process?
2. Why are the withered leaves crushed and cut in the 'rolling' process before fermentation?
3. Do you think uncut and uncrushed leaves would ferment? Explain your answer.
4. Why are the fermented leaves dried in *hot* air?

In brief
Drinks

1. All drinks contain water.
   - Some drinks have water in them already.

   - Other drinks are made by adding water.

2. We all need water to live.
   Communities need large quantities of water for drinking and other uses.
   In some countries, water shortage is a severe problem.

3. Drinking water (tap water) usually contains dissolved substances.

4. Drinks such as tea, lemonade and wine, also contain dissolved substances. In these drinks
   - water is the **solvent**,
   - the dissolved substances, like sugar in tea, are called **solutes**.

   Dissolved substances are described as **soluble**. Substances which do not dissolve are described as **insoluble**.

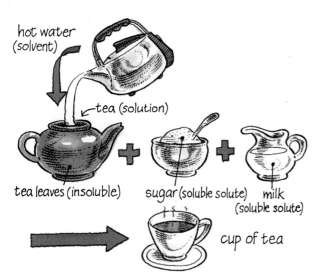

Figure 11 *Making a cup of tea*

5. Water can be made safe to drink by
   - filtering out insoluble impurities like sand and mud,
   - killing bacteria by boiling or by adding chlorine.

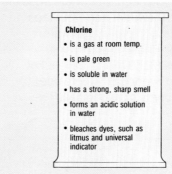

Chlorine
- is a gas at room temp.
- is pale green
- is soluble in water
- has a strong, sharp smell
- forms an acidic solution in water
- bleaches dyes, such as litmus and universal indicator

Figure 12 *Properties of chlorine – a soluble gas used to purify water*

6. Carbon dioxide is added to some drinks to make them fizzy.

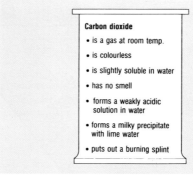

Carbon dioxide
- is a gas at room temp.
- is colourless
- is slightly soluble in water
- has no smell
- forms a weakly acidic solution in water
- forms a milky precipitate with lime water
- puts out a burning splint

Figure 13 *Properties of carbon dioxide*

7. Alcoholic drinks are made by fermenting sugary solutions.

8. Alcoholic drinks can be made stronger by **distillation**.

   distillation = evaporation then condensation

   Different liquids boil at different temperatures. For example, alcohol boils at 78°C and water boils at 100°C. When a mixture containing water and alcohol is heated, alcohol evaporates more easily. When the vapour condenses, the liquid contains a higher percentage of alcohol than the original mixture. This is how spirits like whisky and gin are produced.

9. Although alcoholic drinks are enjoyed by many people, they can result in very serious problems.

10. All substances and materials are made up of **particles**. The preparation and properties of drinks can be explained using the idea of particles (particulate theory).

Thinking about Chemistry and drinks

1 Where does our water come from?

Every time you turn on the tap you expect to get as much water as you need. But where does the water come from? The water from your taps comes via underground pipes from water treatment plants. In England and Wales, 16 200 million litres (3360 million gallons) of water are supplied every day. Where does this treated water come from originally?

Figure 15 *One of the large reservoirs which supply water to people living in the Thames Valley.*

Figure 14 *The water cycle*

Figure 14 shows where water is found on the earth and what happens to it. You can see that it is constantly changing from one form to another in a cycle. Heat from the sun causes surface water to evaporate from rivers, lakes and oceans. This water vapour collects as clouds. As the clouds rise, they cool down. This causes the water vapour to condense as drops of water (or snow flakes).

When it rains, most of the water soaks into the earth as ground water. The rest runs off the land and into streams, rivers and lakes as surface water, much of which seeps back into the oceans.

We take our water from two sources:
a) **surface water** in rivers, lakes and reservoirs (figure 15),
b) **ground water** in underground wells.

Even when surface soil is dry and dusty, porous layers below ground can act like sponges and store vast amounts of water.

2 How is drinking water purified?

Most of the water in our streams and rivers is unsuitable for drinking. It has probably fallen as rain through polluted air, then run along muddy fields or dirty streets. In earlier centuries clean river water was hard to come by and therefore expensive (figures 16 and 17). Nowadays water is cleaned before we use it. This takes place at the waterworks or water treatment plant. Figure 18 opposite is a diagram of a typical water treatment plant. In figure 18 each process is named and its effect is explained.

In some areas, the water is treated further. This often involves the following two processes:
pH adjustment Some waters are acidic enough to react with metal pipes. Lime can be added to neutralise the acid and adjust the pH.
Fluoridation In some areas, about one gram of fluoride is added to every million grams of water (i.e. 1 part per million; 1 ppm). This helps to prevent tooth decay and to reduce bone weaknesses in elderly people.

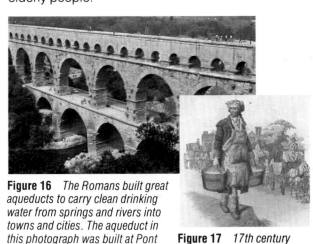

Figure 16 *The Romans built great aqueducts to carry clean drinking water from springs and rivers into towns and cities. The aqueduct in this photograph was built at Pont du Gard in Southern France.*

Figure 17 *17th century seller of clean river water*

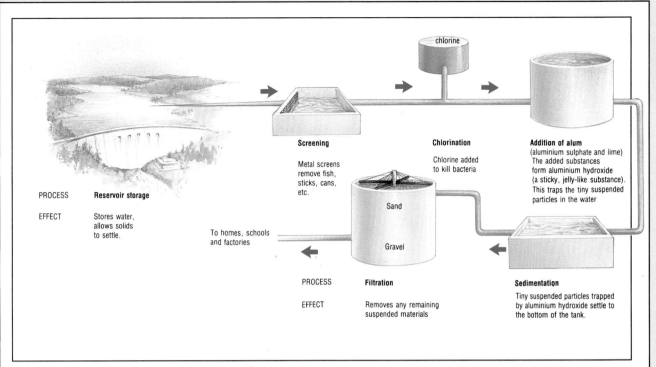

Figure 18 *Processes at a typical water treatment plant*

3 How are alcoholic drinks made?

All alcoholic drinks are made by **fermenting** sugary solutions. People have been fermenting the sugars in honey and fruit juices for at least ten thousand years.

Sugars are carbohydrates. They contain the elements carbon, hydrogen and oxygen. The correct name for table sugar is **sucrose**. This can be used to make alcoholic drinks. Other sugars are also used. These include **maltose** from barley which is used to make whisky. Starch can also be used for alcoholic drinks in place of sugar. This is because it breaks down fairly easily to from maltose. Table 1 shows the source of sugar for different alcoholic drinks.

Table 1 *The source of sugar for different alcoholic drinks*

Drink	Source of sugar	Sugar fermented
Beer	Barley	Starch → maltose
Cider	Apples	Sucrose, maltose and other sugars
Wine	Grapes	Sucrose, maltose and other sugars
Sherry	Grapes	Sucrose, maltose and other sugars
Whisky	Barley	Starch → maltose
Gin	Wheat (corn)	Starch → maltose
Vodka	Potato	Starch → maltose
Saki	Rice	Starch → maltose

Figure 19 *Vat of fermenting lager*

Fermentation needs yeast as well as sugar. Yeast is a micro-organism. The yeast lives on the sugar and splits it up into carbon dioxide and alcohol.

$$\text{sugar} \xrightarrow{\text{yeast}} \text{alcohol} + \text{carbon dioxide}$$
(sucrose; maltose)

Figure 20 shows a simple experiment which you may have used to make a solution of alcohol by fermentation.

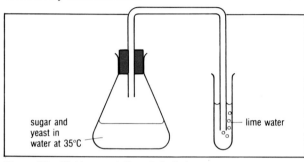

Figure 20 *Making a solution of alcohol*

After about 30 minutes, the lime water turns milky as carbon dioxide is produced. At the same time, alcohol is left in the flask. However, alcohol is a poison for the yeast. When the alcohol concentration gets to about 15 per cent by volume, all the yeast is killed. Because of this, it is impossible to make alcoholic drinks containing more than 15 per cent alcohol by fermentation alone.

There are two ways of making drinks with more than 15 per cent alcohol.

a) **Distilling** fermented liquids. This method is used to make whisky, gin, vodka and saki. These alcoholic drinks are called **spirits** (table 2).
b) Adding extra alcohol. This method is used to make sherry and port. These drinks are called **fortified wines** (table 2).

Table 2 *The percentage of alcohol in different drinks*

Drink	% alcohol	How it is made
Beer	4	fermented barley
Wine	10	fermented sugars in grape juice
Sherry	20	wine with added alcohol
Whisky	40	distilled liquid from fermentation of barley
Gin	40	distilled liquid from fermentation of wheat (corn)
Brandy	40	distilled wine

4 What happens in terms of particles when tea is made?

i) Boiling the water

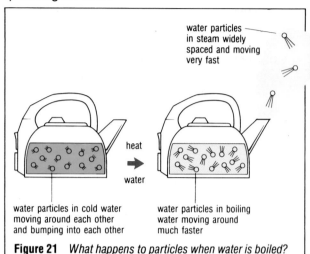

Figure 21 *What happens to particles when water is boiled?*

ii) Brewing the tea

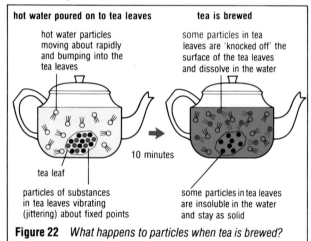

Figure 22 *What happens to particles when tea is brewed?*

iii) Adding milk to the tea

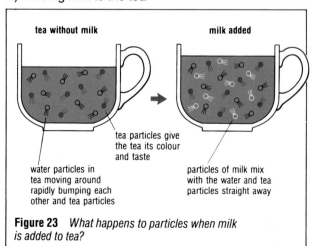

Figure 23 *What happens to particles when milk is added to tea?*

Things to do

Things to try out

1 *Do a survey.*
   a) Carry out a survey of drinks among your friends and your family. Ask them what drinks they have had in the last 24 hours. Make a bar chart or a pie chart of your results.
   b) Carry out a survey of the popularity of different fruit juices among your friends and your family. Your survey could include the following juices: apple, grapefruit, orange, pineapple, tomato. Make a bar chart or a pie chart of your results.

2 *Make a drink from oranges or lemons.*
   a) Plan a recipe for an orange or lemon drink.
   b) Now compare your recipe with one in a cookery book.
   c) What changes will you make to your recipe?
   d) Make the drink.

3 *Collect cuttings related to drinks.*
   a) Collect cuttings related to drinks from magazines or newspapers.
   b) Write six sentences which pick out important points in the cuttings.

4 *Design a poster.*
   Design a poster which could be used as an advert for your favourite drink.

5 *Make tea.*
   a) Plan an experiment to see whether there is any difference between tea made
      A. by putting milk into a cup first and then pouring in the tea or
      B. pouring out the tea and then adding the milk.
      Describe
      i) the experiment itself,
      ii) the care you will take to make the comparisons fair,
      iii) the tests you will use to check whether there is any difference.
   b) Carry out your experiment.
   c) What are your conclusions?

Things to find out

6 Find out about the water supply to your home. Some questions to consider are given below.
   a) How much do you pay in water rates?
   b) Where does your water come from?
   c) Where is the nearest water treatment plant to your home?
   d) What other services does your water company provide besides simply supplying water?
   e) How does your water supply company spend the money you pay in water rates?

7 Use a library or an encyclopaedia to find out about one of the following drinks:
   milk, coffee, fruit juice, beer, whisky.
   Some possible questions to consider are:
   a) How is the drink made or processed before it goes on sale?
   b) What chemicals (constituents) does the drink contain?
   c) What regulations cover the labelling and sale of the drink?
   d) How have sales of the drink changed during the last ten years or so?
   e) What arrangements are there for returning or recycling the containers for the drink?

8 Find out the solute(s) and solvent in the following solutions:
   a) brine   b) vinegar
   c) wine   d) lemonade
   e) tap water   f) liquid paper

Points to discuss

9 Suppose that two of your best friends have got into the habit of drinking heavily. What advice would you like to give them?

10 Some people think it is wrong to have any drinks which contain alcohol. What do you think?

11 Helen who is fifteen years old says, 'It's just being sociable to go to the pub for a drink. If I were the only one having a coke, I'd feel left out.' What do you think?

12 Some people believe that it is more important to improve the water supply and water treatment in poorer countries than to improve the hospital care and education in these countries. What do you think?

Questions to answer

13 Describe the experiments that you would carry out in order to:
   a) find the amount of insoluble solid suspended in 10 cm^3 of a sample of muddy water,
   b) find the amount of solids dissolved in 10 cm^3 of clean river water.

14 Micro-organisms such as yeast can obtain energy from foods such as sugar. If there is no oxygen present, the yeast lives on the sugar producing carbon dioxide and alcohol. This process is called **fermentation**. Two students set up the apparatus in figure 24 at room temperature. They were trying to study the best temperature for fermentation.

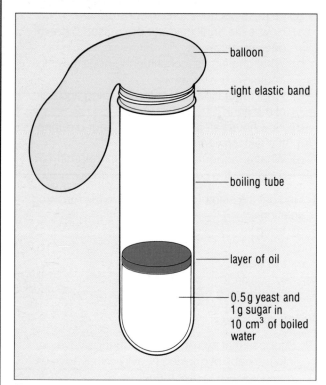

Figure 24

a) What is the balloon for?
b) Why is the tight elastic band used?
c) What happens to the balloon during the experiment?
d) Why did the students use boiled water?
e) Why is the fermenting mixture covered with a layer of oil?
f) Write a word equation for the fermentation process.
g) What should the students do now to find the best temperature for fermentation?

15 Explain the following observations using the idea of particles.
a) Sugar dissolves in water to form a clear solution.
b) Tea has its own special smell.
c) Sugar dissolves more quickly in hot tea than in cold tea.

16 Sugar solutions are often used in cooking. When sugar dissolves in water to form a solution, sugar is the solute and water is the solvent. Figure 25 shows how the percentage of sugar in the solution affects its boiling point.

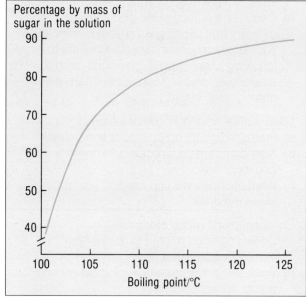

Figure 25

a) What happens to the boiling point of the sugar solution as the percentage of sugar increases?
b) Why do you think the percentage of sugar affects the boiling point?
c) Sugar solutions are boiled to make jam and fondant icing. In jam making, fruit, sugar and water are boiled to 105°C. Then the jam is allowed to set. When fondant icing is made the sugar solution is boiled at 114°C before cooling. What is the concentration of sugar in i) jam, ii) fondant icing?

17 Read section 2, on page 18. *How is drinking water purified?*
a) Make a summary of the stages in purifying our water supplies.
b) In many areas, water is taken from a river *above* a town and used water (**effluent**) is put in the river *below* the town. Why is this?
c) Why do some dentists tell their patients to use fluoride toothpaste?

18 Suppose you are a hospital doctor. One of your patients seems to be urinating a much higher proportion of the liquid that he drinks than is normal. How would you check this?

WARMTH

Introducing warmth

No-one enjoys being cold. In fact being cold can be fatal. Every year we hear of old people who die of cold in the winter. How do we keep ourselves warm?
- We eat food which gives us energy.
- We keep our homes, schools and workplaces warm.
- We try to avoid losing heat unnecessarily.
- We insulate our bodies with clothes and insulate the places where we live.

Figure 1 *In very cold climates special clothing must be worn. How does this protect the body?* ▷

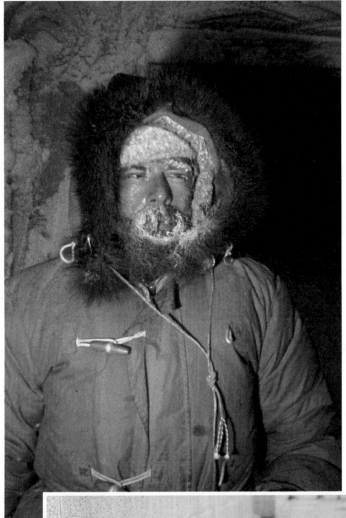

▲ **Figure 2** *What can old people do to keep warm?*

Figure 3 *How do you heat your home? Think of as many ways of heating a home as you can.* ▽

Figure 4 *Marathon runners are sometimes wrapped in aluminium covers after the race to help them conserve energy.* ▷

In this chapter you will see how

- chemistry can help us provide warmth more conveniently and cheaply,
- chemical changes take place when fuels burn,
- our environment is affected when we burn fuels.

Looking at Warmth

1 Heating our homes

We are lucky in Britain because we have plentiful energy supplies. We have **coal**, and **oil** and **gas** from the North Sea. We can use these energy sources to make electricity.

Different people choose different energy sources to warm their homes.

Figure 5 Coal or wood is burned in open fires to provide heating and hot water in the home.

Figure 6 This looks like a coal fire but the fuel is gas – a coal-effect fire.

Figure 7 Electricity is used in electric heaters. Storage heaters use cheaper off-peak electricity.

Figure 8 The oil being drilled in the North Sea is burned in power stations to produce electricity. It can also be burned in domestic boilers to provide cental heating.

Table 1 Information about different energy sources. The costs of the fuels are in pence per megajoule. A megajoule is the amount of energy given out by a small electric fire in about fifteen minutes

Energy source	Cost per megajoule/pence	Advantages	Disadvantages
Coal	0.40	Cheap Coal fires look nice	Dirty Difficult to transport Makes a lot of smoke and ash Difficult to light
Oil	0.48	Easy to transport and store Can be pumped automatically	Needs to be delivered every so often Messy and smelly if it leaks Price is liable to vary
Gas	0.35	Doesn't need to be delivered Can be pumped automatically Cheap Clean	Dangerous if it leaks Supply has to be laid to the house
Electricity	1.38 (0.57 off-peak)	Easily switched on and off Very clean Doesn't need to be delivered	Expensive

Use table 1 to decide the best energy source for each of the people in examples 1 – 3 to heat their homes.

1 Mrs Evans, aged 72, who lives alone in an isolated cottage in Wales. She has only her pension to live on, and she isn't in the best of health.
2 The Patel family, who live in Manchester. The family is four in number, with a good income.
3 The Waddington family, who live in an enormous, draughty house in Yorkshire. They have only a tiny income.

4 Why is electricity so much more expensive than the other energy sources?

2 Designing a gas burner

Many people use natural gas to heat their homes and to do their cooking. You use natural gas as a fuel every time you use a Bunsen burner.

Natural gas is mainly **methane**. Methane contains the elements carbon and hydrogen. When methane burns in plenty of air, carbon dioxide and water are formed.

methane + oxygen → carbon dioxide + water
$CH_4(g)$ + $2O_2(g)$ → $CO_2(g)$ + $2H_2O(l)$

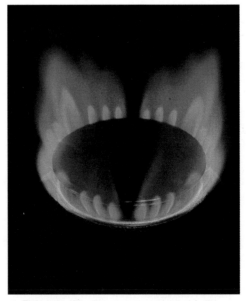

▲ **Figure 9** The flame on a gas cooker is like a Bunsen flame.

◀ **Figure 10** What kind of Bunsen flame is shown here?

If there is a shortage of air, there is not enough oxygen to form carbon dioxide. Poisonous carbon monoxide (CO) may be formed instead. If the air supply is really bad, the carbon may not oxidise at all. It may just form soot. This soot will be deposited on whatever you are heating.

What does an efficient gas burner need?
An efficient gas burner needs two features.

◆ A controlled flow of gas which can be turned on and off and adjusted.
◆ A good supply of air.

Figure 11a shows how a Bunsen burner gets its air supply when the air hole is open. The flame is said to be **aerated**. Figure 11b shows how it gets it with the air hole closed. The flame is **non-aerated**.

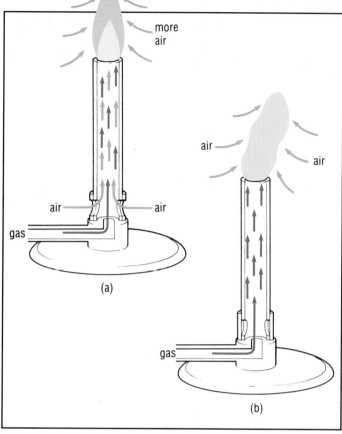

Figure 11
a) Air hole open. Aerated flame: gas already mixed with air when it burns.
b) Air hole closed. Non-aerated flame: gas gets air supply from air round flame.

1 What colour is the flame when the air hole of a Bunsen burner is
  a) open, b) closed?
2 When you are heating things with a Bunsen burner, you should do it with the air hole open. Think of *two* disadvantages of heating things with the air hole closed.
3 Think about the flame on a gas cooker. Is it aerated or non-aerated? How do you know?

The design of gas cooker burners
A gas cooker burner needs to give a clean flame so that it doesn't make saucepans sooty on the outside. The flame must be easily controlled. Figure 12 shows the design of a typical burner.

4 In what ways is this burner similar to a Bunsen? In what ways is it different?

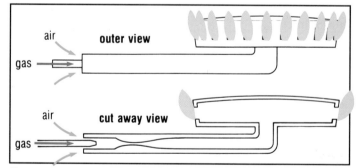

Figure 12 *A typical gas burner*

Design a burner
The Northern Glass Company makes glass vases. One of the steps in the manufacture of a vase involves heating a thick solid glass rod evenly (figure 13).

Gas is used to heat the rod.
Sketch the design of a burner that would do this job.

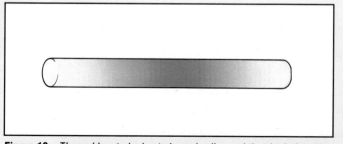

Figure 13 *The rod has to be heated evenly all round the shaded region.*

3 Fire!

Home fires kill about 800 people each year in Britain. Fires are often easy to light, but difficult to put out.

The fire triangle
A fire needs three things:

◆ fuel – anything that will burn,
◆ oxygen – usually from the air,
◆ a high temperature to start the fire and keep it going.

These three things make up the **Fire Triangle** (figure 14). You can put out a fire by taking away any one of the three.

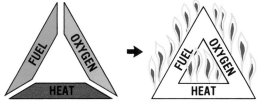

Figure 14 *The Fire Triangle*

Figure 15 *What fire-fighting methods would be used to fight a fire like this?*

How to put out fires
Table 2 gives some of the methods that can be used to put out fires.

Table 2 *Fire-fighting methods*

Fire-fighting method	What it involves	Where you get it
Fire blanket	A blanket made of non-flammable material is thrown over the fire	Red-painted fire blanket container
Sand	Sand is thrown on the fire	Red-painted fire bucket
Water	Cold water is poured over the fire. Cannot be used on fires involving oil, petrol or electricity	Tap, fire hose or red-painted water-type fire extinguisher
Carbon dioxide	Carbon dioxide gas is directed at the fire, and forms an invisible blanket over it	Black-painted carbon dioxide-type fire extinguishers
Foam	A foam containing bubbles of carbon dioxide is poured over the fire	Cream-painted foam-type fire extinguishers
Halon	A non-flammable liquid made from carbon, chlorine, bromine and fluorine is poured on the fire. It forms a dense blanket of vapour	Green-painted halon-type fire extinguishers
Powder	A fine powder is poured over the fire	Blue-painted powder-type fire extinguishers

Use table 2 to answer these questions.
1. Each of the fire-fighting methods in the table works by taking away one or more of the three sides of the Fire Triangle. For each method, decide how it works.
2. Why must water not be used against oil, petrol or electrical fires?
3. Look around your chemistry laboratory at school. List the different kinds of fire-fighting equipment in the laboratory.
4. What would you do in each of the following situations?
   a) You are having a bonfire in the garden. The fire gets out of control and the hedge starts to burn.
   b) You are cooking chips when the chip pan catches fire.
   c) You are babysitting when a small girl's nightdress catches fire.
   d) You are in the laboratory studying the Warmth chapter. You are testing the value for money of methylated spirit as a fuel. Suddenly the boy next to you spills a bottle of methylated spirit. It spreads over the bench and catches light.

4 Cleaning up the power stations

One of the worst **air pollutants** is **sulphur dioxide**, (SO_2). Chemists believe that sulphur dioxide is one of the major causes of **acid rain**. Acid rain can easily damage plants and animal life as well as buildings. Coal-fired power stations are major sources of sulphur dioxide poisoning. These power stations burn coal to generate electricity. Coal contains about 1.5 per cent sulphur. The sulphur is in the form of compounds. When coal burns, the sulphur burns too, forming sulphur dioxide. Nearly half of the sulphur dioxide given out by human activities each year comes from power stations.

The people who run the power stations are trying to find ways of cutting down the amount of sulphur dioxide given out. They have looked at several possible solutions. These include:

- Using coal with less sulphur in it. This would have to come from abroad.
- Building nuclear power stations to replace coal-fired stations.
- Getting the sulphur out of the coal *before* burning it.
- Getting the sulphur dioxide out of the gases *after* burning the coal.

In fact they chose the last solution – getting the sulphur dioxide out of the gases after burning the coal. This is called **flue gas desulphurisation** (FGD) (figure 17).

How does FGD work?

Sulphur dioxide is an acidic gas. Like all acids, it is neutralised by alkalis. The alkali used is lime – calcium oxide (CaO). This is a very cheap alkali which is made from limestone.

The acidic sulphur dioxide reacts with the alkaline lime. This forms calcium sulphite ($CaSO_3$), which is a solid.

sulphur dioxide	+	calcium oxide (lime)	→	calcium sulphate
$SO_2(g)$	+	$CaO(s)$	→	$CaSO_3(s)$

The calcium sulphite is then reacted with air. If forms calcium sulphate ($CaSO_4$), which is also called **gypsum**.

calcium sulphite	+	oxygen (air)	→	calcium sulphate
$2CaSO_3(s)$	+	$O_2(g)$	→	$2CaSO_4(s)$

Figure 16 *Fitting a flue gas desulphurisation plant is expensive. For a big power station like this it would cost about £200 million to fit. It would also cost about £30 million a year to run.*

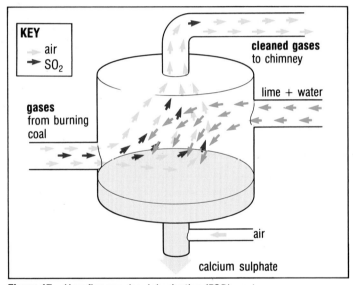

Figure 17 *How flue gas desulphurisation (FGD) works*

Calcium sulphate is a useful chemical. It is used to make plaster. Plaster is important in the building trade.

1. Suppose all Britain's coal-fired power stations were fitted with FGD. They would produce about 8 million tonnes of calcium sulphate a year. The building industry needs about 3 million tonnes of calcium sulphate a year. What would happen to the rest?

2. Where would the money for fitting FGD come from? Who would pay for it? Do you think it is worth the expense?

In brief
Warmth

1. Fuels burn in air, giving out heat. This is called combustion. A reaction which gives out heat is called an **exothermic** reaction.

2. Combustion needs three things: fuel, oxygen and heat. The rate of combustion depends on the conditions, particularly the concentration of oxygen. Fuels burn much faster in pure oxygen. Sometimes they burn so fast that they explode. (There is more on combustion in the *Thinking About* section.)

3. Combustion is a chemical reaction. A chemical reaction always involves the formation of a new substance. Fuel reacts with oxygen to form new substances. Most fuels contain carbon and hydrogen, and these combine with oxygen to form carbon dioxide and water when the fuel burns. This process is called **oxidation**. Fuels may also form other, unwanted substances when they burn (figure 18).

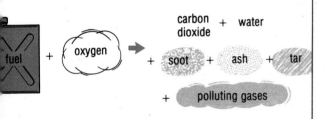

Figure 18 *Products of burning fuel*

4. Different fuels have different properties. Figure 19 shows some of them.

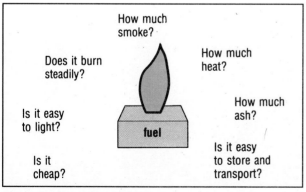

Figure 19 *Properties of a fuel*

5. Coal burns to form soot, ash, tar and acidic gases.

6. Different fuels are suitable for different purposes. We choose a fuel for a particular job according to its properties. Cost, convenience and pollution are particularly important when deciding which fuel to use. No fuel is completely ideal.

7. Fuels are often processed to improve them. Coal is processed to make smokeless fuels such as **coke**. Crude oil is processed to give **petrol**, **paraffin** and other fuels.

8. Pollution is the contamination of the environment by substances made by humans. Combustion of fuels often causes air pollution.

9. Figure 20 shows some of the pollutants that may be formed when fuels burn. The amounts of pollutants that are formed depend on the fuel, and the way it is burned.

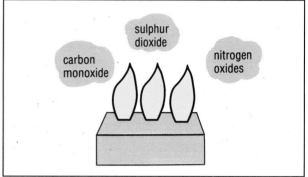

Figure 20 *Some common pollutants formed when fuels burn*

10. Many fuels, particularly coal and coke, contain sulphur. When they burn, the acidic gas sulphur dioxide is formed. This can help form **acid rain**, as figure 21 shows.

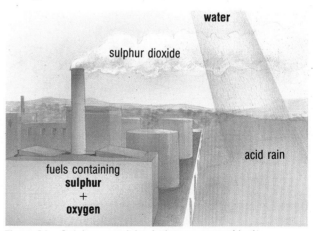

Figure 21 *Sulphur-containing fuels can cause acid rain*

11. Acid rain is harmful to living things. It is harmful to trees, and to life in rivers and lakes. Acid rain and sulphur dioxide in the air cause damage to buildings. They make metals corrode faster.

12. Pollution can be cut down, but this costs money. We need to balance the cost against the benefits of controlling pollution.

Thinking about Chemistry and warmth

1 What is energy?

We depend on energy to keep things going. Without energy, life would grind to a halt.

Energy comes in different forms. Figure 22 lists some important forms of energy.

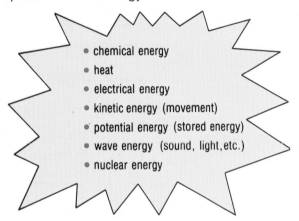

- chemical energy
- heat
- electrical energy
- kinetic energy (movement)
- potential energy (stored energy)
- wave energy (sound, light, etc.)
- nuclear energy

Figure 22 *Important forms of energy*

Energy can be converted from one form to another. Figure 23 shows the energy conversions which go on in a coal-fired power station.

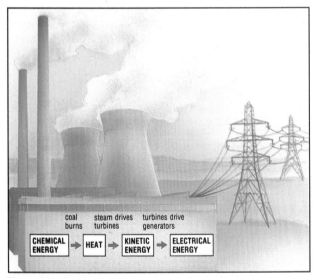

coal burns → steam drives turbines → turbines drive generators

CHEMICAL ENERGY → HEAT → KINETIC ENERGY → ELECTRICAL ENERGY

Figure 23 *Energy conversions in a power station*

Where do we get energy from?
Society needs energy sources to keep things going. The most important energy sources are **fuels**. Fuels contain stored chemical energy. They burn in air, releasing the chemical energy as heat. This heat can be used to keep us warm, or to drive motor vehicles. It can drive power stations which make electricity.

Food is a fuel. We need it to provide the energy to keep our bodies going.

2 What happens when fuels burn?

Another name for burning is **combustion**. During combustion, a fuel reacts with oxygen. This oxygen usually comes from the air.

Most fuels contain the elements carbon and hydrogen. When the fuel burns, the carbon and hydrogen are oxidised. They form carbon dioxide and water. For example, with petrol:

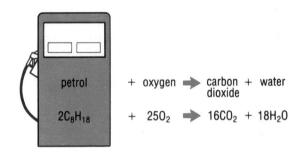

petrol + oxygen → carbon dioxide + water

$2C_8H_{18}$ + $25O_2$ → $16CO_2$ + $18H_2O$

Air is only about one-fifth oxygen. The rest is mostly unreactive nitrogen. Only the oxygen is involved in combustion. The nitrogen does not react.

Fuels burn much more fiercely in pure oxygen than they do in air. Figure 24 shows an oxy-acetylene burner. Acetylene is a gas which burns in air, rather like natural gas. But in pure oxygen it burns much more fiercely. An oxy-acetylene flame is hot enough to cut through steel.

Figure 24 *Oxy-acetylene torches are used for cutting through steel*

3 What happens when there isn't enough oxygen?

Sometimes fuels cannot get all the oxygen they need to burn. For example, in a car engine the petrol may not get enough air to burn properly.

When the oxygen supply is poor, fuels burn to give different products. With less oxygen available, the carbon in the fuel forms carbon monoxide instead of carbon dioxide. Carbon *monoxide* contains only half as much oxygen as carbon *dioxide*.

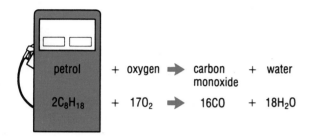

If the oxygen supply is really bad, there may not even be enough to form carbon monoxide. The carbon stays unchanged and unoxidised. It comes off as black, sooty smoke.

Carbon monoxide is a dangerous gas, because it is very poisonous. It stops the blood carrying oxygen properly. Your body can cope with small amounts of carbon monoxide, but large amounts can kill.

Cigarette smoke contains a lot of carbon monoxide (figure 25). This is one of the many reasons why smoking is so bad for your health.

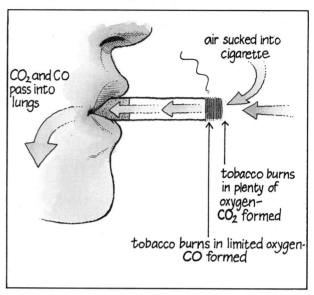

Figure 25 *Why cigarette smoke contains carbon monoxide*

Carbon monoxide is also present in car exhaust gases. It may also be given off by gas burners and paraffin heaters which have their air supply blocked with dirt.

Taking it further

To survive, your body must have a constant supply of oxygen. Oxygen is carried around the body in the blood. Blood contains a substance called **haemoglobin**. Haemoglobin combines with oxygen, but the bond between them is weak.

When blood passes into the lungs, lots of haemoglobin molecules form weak bonds to oxygen molecules, forming **oxyhaemoglobin**. When the blood arrives in other parts of the body, where oxygen is in short supply, the oxyhaemoglobin breaks down. Haemoglobin is reformed, and oxygen is released. The haemoglobin is now free to pick up another oxygen molecule in the lungs (figure 26a).

The bad news is that carbon monoxide can also combine with haemoglobin, to form **carboxyhaemoglobin**. And the bond between carbon monoxide and haemoglobin is *strong*. So carbon monoxide cannot be easily removed from the haemoglobin. This stops the haemoglobin being

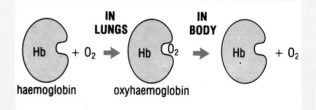

Figure 26 a) *Haemoglobin and oxygen*

Figure 26 b) *Haemoglobin and carbon monoxide*

able to carry oxygen. If too many haemoglobin molecules get blocked by carbon monoxide, the blood cannot carry enough oxygen to keep the body going.

4 Why do burning fuels cause pollution?

Fuels produce gases when they burn. The main gases are

- Carbon dioxide (CO_2) formed when carbon in the fuel is oxidised.
- Sulphur dioxide (SO_2) formed when sulphur impurities in the fuel are oxidised.
- Nitrogen oxides (NO and NO_2) formed when nitrogen and oxygen in the air combine together.

All chemical elements can be classed as metals or non-metals. Metals have very different properties from non-metals. Their oxides have very different properties too. In particular

- Metal oxides are always solids.
  Non-metal oxides are often gases or liquids.
- Metal oxides are alkaline.
  Non-metal oxides are acidic.

You can see why non-metal oxides like SO_2 and NO_2 cause such problems when fuels are burned. Not only are these oxides acidic. They are also gases, so they escape into the air and cause pollution.

Why do acid gases cause damage?
Acid gases are most damaging when they combine with rain water. They react, forming acidic solutions.

- Sulphur dioxide forms sulphurous acid (H_2SO_3) and sulphuric acid (H_2SO_4).
- Nitrogen oxides form nitric acid (HNO_3).

These acidic solutions are only very dilute, but they are still very corrosive. They attack stone and metal, and are harmful to plant and animal life.

Figure 27 *Non-metal oxides escape into the air from the chimney of a coal-fired power station*

Taking it further

Figure 28 shows the types of oxides formed by elements in different parts of the Periodic Table.

1. Whereabouts in the Periodic Table are metals to be found?
2. Whereabouts are non-metals to be found?
3. What kind of elements are found in the middle of the table?

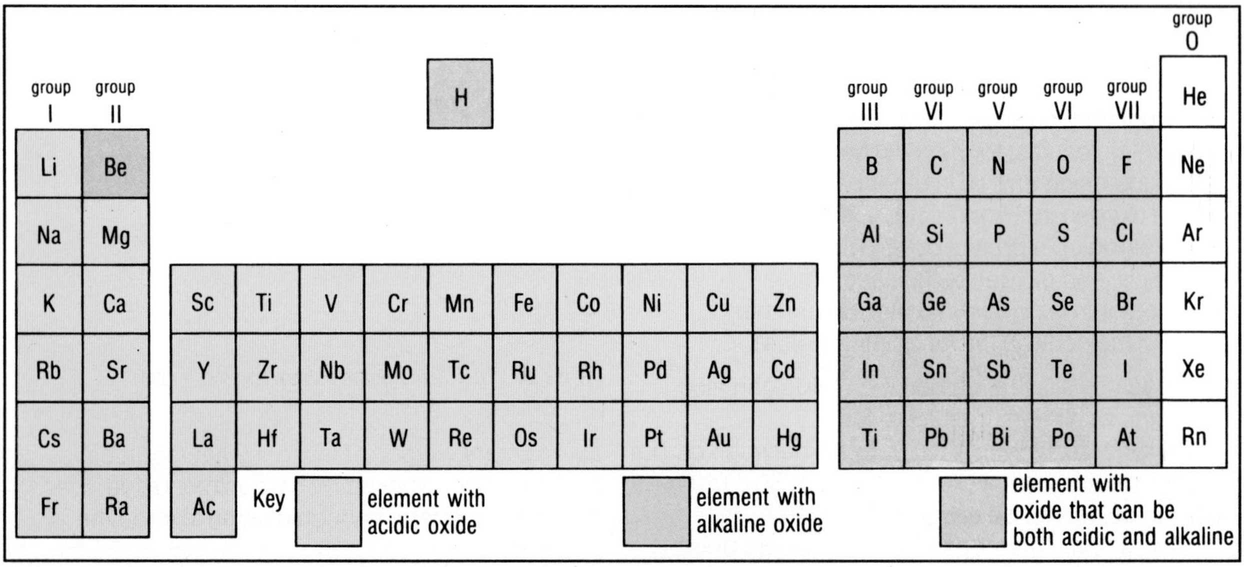

group I	group II										group III	group VI	group V	group VI	group VII	group 0
				H												He
Li	Be										B	C	N	O	F	Ne
Na	Mg										Al	Si	P	S	Cl	Ar
K	Ca	Sc	Ti	V	Cr	Mn	Fe	Co	Ni	Cu	Zn	Ga	Ge	As	Se	Br	Kr
Rb	Sr	Y	Zr	Nb	Mo	Tc	Ru	Rh	Pd	Ag	Cd	In	Sn	Sb	Te	I	Xe
Cs	Ba	La	Hf	Ta	W	Re	Os	Ir	Pt	Au	Hg	Ti	Pb	Bi	Po	At	Rn
Fr	Ra	Ac														

Key: element with acidic oxide / element with alkaline oxide / element with oxide that can be both acidic and alkaline

Figure 28 *Types of oxides formed by elements in different parts of the Periodic Table*

Things to do

Things to try out

1 *Were they the good old days?*
   For this activity you will need to talk to a person who is over 70 years old.

   Ask the person what things were like when they were your age. Ask them
   a) What fuel was used to warm their home?
   b) Was every room heated? If not, which rooms were heated?
   c) Which fuel was used for cooking food?
   d) What means of transport did they use:
      i) for journeys less than 2 miles or so,
      ii) for journeys over 2 miles or so?

2 *Making a camp fire*
   You're out on a picnic, but you've forgotten the camping stove. It is a windy day. You have to make a fire to boil the kettle. You can use only the materials shown in figure 29. You don't necessarily have to use all of them.

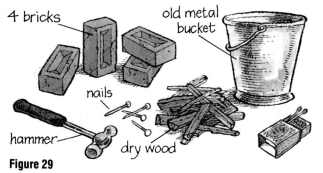

Figure 29

   Sketch your camp fire design. Briefly explain why you chose this design.

Things to find out

3 What is self-warming food? How does it work?

4 The fuels listed below are not very common in Britain. For as many as you can, find out what they contain, and where they are used.
   a) Charcoal b) Peat c) 'Meta' fuel
   d) Bagasse

5 Petrol, paraffin, diesel oil and fuel oil are all made from crude oil. How?

6 Some people use solar panels. How do they work?

7 North Sea Gas has only been in use since the 1970s. What gaseous fuel was used before then? Where did it come from?

Making decisions

8 *Energy sources for industry*
   Look back at table 1 on page 25. Which energy source would be best for each of the following? Give a reason for each choice.
   a) Fuelling a power station in Yorkshire.
   b) Fuelling a power station in Saudi Arabia.
   c) Heating an oven in a large bakery.
   d) Heating steel bars in a steelworks, before rolling them to make steel plate.
   e) Heating a large greenhouse used for growing tomatoes.

Points to discuss

9 It's a winter evening and you're sitting watching television. You're wearing a T-shirt and jeans. The heating is on, but even so you begin to feel a bit chilly. Do you
   a) put on a pullover, or
   b) turn up the heating?
   Which is the most sensible choice? Why?

10 In winter, many old people in Britain suffer from the cold. They may even develop a condition called hypothermia. They get so cold their bodies stop working properly.
   Why do you think old people are particularly likely to suffer from the cold?

11 Suppose you are Minister for the Environment. What laws would you pass to try and cut down air pollution? Remember – laws have to be *enforceable*. You have to be able to prove they are being broken. Remember too that it is impossible to remove air pollution *completely*.

12 Controlling acid rain could put up the price of electricity in Britain. Why? Do you think British people would be prepared to pay?

13 CHEMCO is a company which makes chemicals in a small town in the Midlands. CHEMCO employs 400 local people.
   CHEMCO is in trouble because the factory is giving out polluting gases. Local residents have complained about the smell. They say pollution is attacking their houses and spoiling their garden crops.
   CHEMCO says that to control the pollution would cost £20 million. They could not afford this without laying off one tenth of the workforce.

   Discuss what you think the local residents should do.

Questions to answer

14 Copy the following paragraph and fill in the missing words or groups of words. Each word or group of words is used only once. The missing words and groups of words are: **carbon dioxide, carbon monoxide, combustion, exothermic, oxides of nitrogen, sulphur dioxide, water.**

Fuels are substances which burn in air, giving out heat. Another name for burning is __(a)__. Reactions which give out heat are described as __(b)__. Most fuels contain the elements carbon and hydrogen. When the fuel burns in plenty of air, the carbon in it forms __(c)__. But if the air supply is limited, __(d)__ is formed instead. The hydrogen in the fuel forms __(e)__ when it burns. Burning fuels can cause air pollution. Common pollutants are __(f)__ and __(g)__.

15 Three plastic bottles were filled with different gases, as shown in figure 30.

Figure 30

The three bottles were put behind a safety screen. For each bottle in turn, the stopper was removed and a long burning taper was held over the open mouth.

a) This experiment must be carried out behind a safety screen. Why?
b) What would you expect to happen with each bottle?
c) Explain why the gas in each bottle behaved in this way.
d) A gas leak in a house can be very dangerous. Explain why.

16 You have been given a white solid material. You are told it is a fuel, and that it costs about the same as coal.
a) What properties would you want the solid to have in order to be a good fuel?
b) What tests would you do to decide whether it had these properties?

17 Sharon did an experiment to compare wood and charcoal as fuels. She wanted to know which fuel caused least air pollution. The apparatus she used is shown in figure 31.

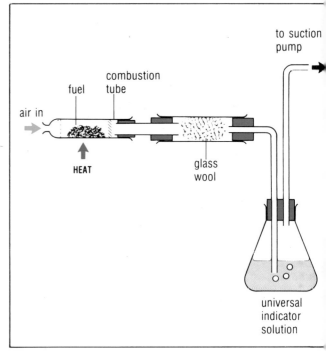

Figure 31 *Apparatus for investigating the air pollution caused by different fuels*

Her results are shown in table 3.

Table 3

Fuel	Appearance of glass wool afterwards	Colour of indicator solution afterwards
Wood	yellow-brown, tarry	yellow (pH 6)
Charcoal	clean	yellow (pH 6)

a) Describe how the apparatus works. Your description should begin 'Air is drawn into the combustion tube ...'
b) What does the experiment tell you about wood as a fuel, compared with charcoal?
c) What other things would you need to know about wood and charcoal before you could decide which was the better fuel?

CLOTHING

Introducing clothing

People wear clothes for all sorts of reasons – to keep warm, to keep cool, to keep dry – and to look good! Clothes are usually made up from pieces of fabric, which is another name for cloth. Fabric is built up from thin strands, or fibres, like cotton, wool, nylon and polyester woven together.

Figure 1 *What are the advantages and disadvantages of metal clothing? Apart from knights, who else might wear metal clothing?* ▶

Chemists help to make better clothing for you. They design new fibres, and improve old ones. They find ways to treat fabrics, to make them waterproof for example. They also make dyes to give fabrics attractive colours. What properties do clothing fabrics need? Look at the captions for figures 1 – 4 and think about the questions.

Figure 2 *What are the advantages and disadvantages of paper clothing?* ▶

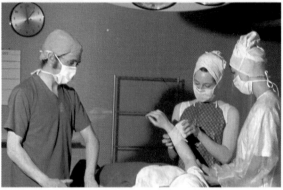

Figure 3
Think of five useful properties of the clothing fabrics you are wearing at the moment. ▼

Figure 4
What are the advantages and disadvantages of plastic clothing? ▶

In your answers to these questions, you probably thought about the properties of the materials involved. For example, you know that metals feel cold and paper burns.

In this chapter you will see how

- ◆ chemistry can help us understand the properties of clothing materials,
- ◆ chemists can use their skills and knowledge to design even better materials.

Looking at Clothing

1 Dyes for brighter clothing

Most fibres are naturally white, or else a rather dull colour. Dyes are used to give bright colours to clothing.

Dyes are coloured substances. But any old coloured substance will not do – the colour must stick to the cloth. For centuries, people used natural dyes. For example, a blue dye was made from the indigo plant. But the colours of natural dyes are often dull and the range of colours is limited. Also, natural dyes are not **fast** – they fade when repeatedly washed or exposed to sunlight.

Nowadays there is a huge range of synthetic dyes, of any colour you can imagine. What's more, modern synthetic dyes are fast.

In 1856, William Perkin was trying to find a way of making the drug quinine, which is used to treat malaria. Instead of quinine he got a beautiful purple dye, which he called **mauve**. Perkin found that mauve dyed cloth, and was fast to light. The dye became famous and very fashionable.

Perkin became rich because of his discovery. Soon many other synthetic dyes of different colours were discovered by chemists. Chemists learnt more about the chemistry of dyes. This made it possible to produce new dyes on purpose instead of by accident. Nowadays, chemists can make practically any colour dye.

Making dyes stick

For a dye to be fast and not wash out, it must stick to the cloth. Like all substances, dyes contain tiny particles. These particles are attracted to the fibres and stick to them, as shown in figure 9.

▲ **Figure 5** Many dyes have been used to colour these fabrics.

◄ **Figure 6** Indigo is used to dye blue jeans. But the blue fades with washing and sunlight.

◄ **Figure 7** The first synthetic dye was made by an English chemist, William Perkin. He made it by mistake in 1856. He was just 18 years old at the time.

Figure 8 Perkin's dye called 'mauve' was celebrated in the Penny mauve stamp of 1881. ▼

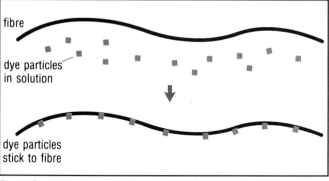

Figure 9

A particular dye is usually only attracted to a particular type of fibre. This means different fibres need different dyes (figure 10).

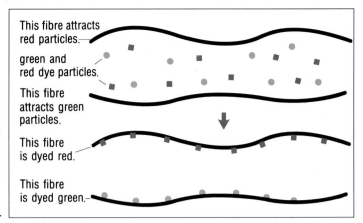

1 Why do you think some dyes are fast while others run?
2 When you wash a brightly-dyed garment for the first time, it is best to wash it on its own, instead of with other clothes. Why?

Figure 10 *Different dyes stick to different fibres.*

2 Dry-cleaning

Some clothes cannot be washed in water. This is because they would shrink or change shape in water. These clothes are labelled to show they have to be dry-cleaned (figure 11). Dry-cleaning is also used to remove dirt and stains that water and detergents cannot remove.

◀ **Figure 11** *The symbols used on a garment label show how it should be cleaned – it must be ironed with a cool iron, should not be bleached and should be dry cleaned.*

Dry-cleaning uses special solvents. These solvents are particularly good at dissolving grease. But they have to be non-flammable – they must not burn, because this could cause a dangerous accident. The solvents must not be poisonous, otherwise they could harm the dry-cleaning workers or the owners of the clothes.

A dry-cleaning machine is really like a big washing machine, but it uses a solvent instead of water. The solvent it uses is recycled and used again and again. It is recycled by distilling. Dirty solvent is boiled, then its vapour is condensed by cooling, to give clean solvent. The dirt is left behind as a solid and can be thrown away.

The fact that the solvent has to be distilled means it needs another property. Its boiling point must not be too high, otherwise it would be difficult and costly to boil. On the other hand, the boiling point cannot be too low, otherwise it would vaporise in the machine.

Chemists can 'design' dry-cleaning solvents so they have exactly the properties that are needed. A solvent that is often used contains the elements carbon and chlorine. It is called tetrachloroethane. It is a colourless liquid, boiling point 121°C.

Figure 12 *Dirty solvent being removed from a dry cleaning machine. It is then purified for re-use.*

1 Petrol is good at dissolving grease. Why is it not used as a dry-cleaning solvent?
2 Is dry-cleaning really 'dry'? Explain your answer.
3 When you take clothes home from the dry-cleaners in your car, you are advised to keep the car windows open. Why?

3 Protective clothes for firemen

Firemen's protective clothes obviously have to be **fireproof**. They also have to be good insulators, to stop heat reaching the body. They have to be waterproof because of all the water that gets pumped onto fires. They have to be durable so they wear well, and also have to be comfortable to wear (figure 13).

It is difficult to get clothing that is fireproof, insulating *and* waterproof. Some plastics *could* be used, but they would melt and stick to the skin at high temperatures.

For many years, firemen wore heavy woollen tunics. As you may know if you have done tests on fibres, wool does not burn easily. Firemen's tunics used to be made of thick, dense wool. The wool was matted so closely that it insulated the body from heat. It was also difficult for water to penetrate. But even so, the wool absorbed a lot of water. Once wet, these jackets became uncomfortably heavy to wear.

So fibre chemists looked for new answers. They found two. One used a natural fibre, one a synthetic fibre.

1 Natural fibre: specially treated wool

The new, lighter weight firemen's tunics are made in three layers as shown in figure 14.

The outer layer is wool, specially treated to make it fireproof and waterproof. The fireproofing treatment is a compound of the element zirconium. The waterproofing treatment uses a fluorocarbon – a non-flammable, water-repellent compound containing carbon and fluorine.

The middle layer is dense, matted wool – as in the traditional jacket, but thinner. It insulates the body against heat. The inner layer is cotton, for comfort next to the skin.

2 Synthetic fibre

In recent years, chemists have developed a very strong, fire-resistant fibre called aramid fibre. It is quite like nylon, but stronger, harder to melt and more difficult to burn (figure 15). Aramid fibre is so strong it can be used to make bullet-proof vests. The photograph shows how fire-resistant it is. It is also water-resistant and comfortable to wear, so it is excellent for making firemen's clothes.

Figure 13 *Firemen's clothes have to protect them from fire, heat and water.*

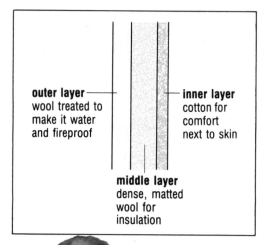

Figure 14 *Three-layer construction of a lightweight woollen fireman's jacket*

Figure 15 *Fire-resistant material being tested in a laboratory.*

The table below gives some information about three fire-resistant fabrics. Fire-resistant cotton and fire-resistant wool are natural fibres, specially treated to make them fire-resistant. Aramid is a synthetic fibre specially designed to be fire-resistant.

1. You are responsible for supplying protective clothing for your local fire brigade. The safety of your firemen is vital, but you also have to consider cost. Which fabric would you choose, and why?
2. What other uses, apart from firemen's clothing, can you think of for fire-resistant fibres?

Table 1

	Fire-resistant cotton	Fire-resistant wool	Synthetic aramid fibre
Fire-resistance	medium/good	good	very good
Insulation	medium/good	good	very good
Durability (how well it wears)	good	good	very good
Comfort	very good	very good	good
Cost	medium	medium/high	high

4 Disposable nappies

A good disposable nappy needs some important properties. It should be
- Absorbent — to soak up lots of liquid
- Leak-proof — so liquid can't leak out at the legs or waistband
- Comfortable — for the baby to wear
- Disposable — the nappy may be disposed of by burning or breakdown by bacteria. As much of the nappy as possible should be biodegradable.

Figure 17 shows a typical disposable nappy.

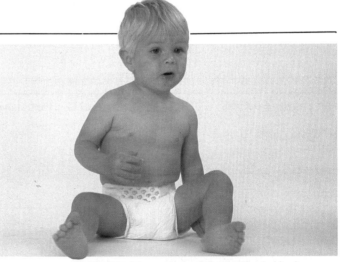

Figure 16 There are around two million babies in Britain. About half of them wear disposable nappies - five per day on average. That makes five million disposable nappies used every day!

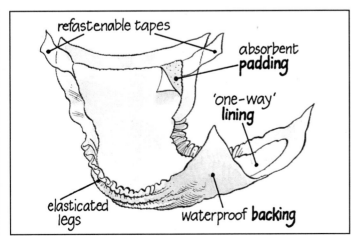

Figure 17 A typical disposable nappy

Chemistry plays an important part in deciding the best material for making the different parts of the disposable nappy. Table 2 on the next page shows some of the materials that might be used.

Water absorption is a particularly important property. Some materials can absorb a lot of water, while some are waterproof and absorb no water at all.

Cotton and tissue paper absorb water well. They are mostly made of cellulose, and cellulose has lots of 'water hooks'.

'Superabsorbent polymer' was specially developed by chemists to absorb large amounts of water. It has large numbers of 'water hooks' and it forms a kind of jelly which holds the water under pressure.

Imagine you are a disposable nappy manufacturer. Use table 2 to answer these questions.

1. Which material or materials would you choose for the backing? Give reasons for your choice.
2. Which material or materials would you choose for the padding? Give reasons for your choice.
3. Which material or materials would you choose for the lining? Give reasons for your choice.
4. Are there any materials you would like to use but which are not in the table?

Figure 18 *This magnified photograph of tissue paper shows how the fibres are bonded together, not woven as in cloth.*

Table 2 *Materials that might be used to make disposable nappies*

Material	Description	Price	Biodegradable?	Flammable?	Waterproof?	Water absorbent?	Strength	Softness
Non-woven polypropylene fabric	see figure 18	low	no	yes	no	no	strong	soft
Non-woven rayon fabric	see figure 18	low	yes	yes	no	no	strong	soft
Woven cotton cloth	white cloth like sheets	high	yes	yes	no	fairly	strong	soft
Aluminium foil	shiny foil	medium	no	no	yes	no	strong	hard
Clear polythene sheeting	clear plastic sheet	low	no	yes	yes	no	strong	soft
Whitened polythene sheeting	opaque, white plastic sheet	low	no	yes	yes	no	strong	soft
Fluffed cellulose pulp (made from wood)	like cotton wool	low	yes	yes	no	yes	weak	soft
Superabsorbent polymer	like cotton wool	high	yes	yes	no	very	weak	soft
Tissue paper		low	yes	yes	no	yes	weak	soft

In brief
Clothing

1. **Clothes**, **fabrics**, **threads** and **fibres** are all important in this chapter. Figure 19 summarises the differences between them.

2. Fibres can be natural or made artificially by chemists. Natural fibres may come from plants, like cotton. They may come from animals, like wool. Fibres made by chemists include nylon and polyester. Figure 20 sums it up.

3. There are many different types of fibres used to make clothes. Wool, cotton, acrylic, polyester and nylon are the most common fibres used. Fibres can be identified by carrying out simple laboratory tests on a piece of fabric made from the fibre. A useful test is to see how the fabric behaves when it is heated.

4. Like all substances, fibres are made up of tiny **particles**. These particles are much too small to see. The properties of a fibre (strength, stretchiness and so on) are decided by what its particles are like.

5. All fibres are polymers. **Polymers** consist of very large particles which are made up of many small particles called **monomers**. The process by which monomer particles join to a polymer particle is called **polymerisation**.

6. The main problem with cleaning clothes is removing grease and oily material. Grease does not dissolve in water. Sometimes special solvents are used instead of water – this is called **dry-cleaning**. But water *can* be used to remove grease, if detergents are added to the water.

7. For wet clothes to become dry, water must evaporate. Evaporation involves the change of a liquid to a gas or vapour. Water particles must escape from the clothes. Some fibres attract water particles more than others. Fibres that do not attract water particles well are easy to drip-dry. But they feel 'sweaty' to wear, because they do not soak up perspiration.

8. Chemists have developed ways of treating fabrics to improve their properties. Figure 21 sums up some of the treatments.

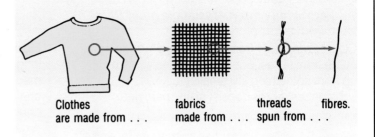

Figure 19 *Clothes, fabrics, threads and fibres*

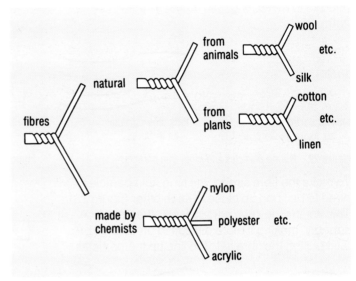

Figure 20 *Different types of fibres*

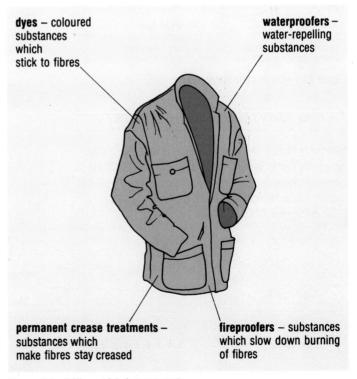

Figure 21 *Different fabric treatments*

Thinking about Chemistry and clothing

1 What kind of particles make good fibres?

Clothes are made from fabrics, and fabrics are made from fibres. But fibres themselves are made up from something still smaller. Like all substances, fibres consist of tiny particles.

Fibres are long and thin, and they need to be strong. So while substances such as water and air contain particles which are roundish in shape, this would not do for fibres. Long, thin particles are needed to make long, thin fibres, as shown in figure 22.

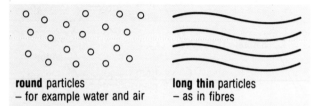

Figure 22 *Round particles and long, thin particles*

To make the fibre strong, the long, thin particles need to be lined up close to each other (figure 23). This means they can attract each other more strongly, making it more difficult to break the fibre. Just pulling the fibre helps to line up the particles.

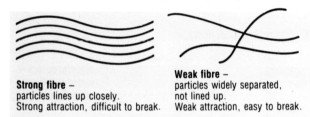

Figure 23 *How to get a strong fibre*

2 How can you make stretchy fibres?

Stretchy fibres like nylon or wool need to contain 'stretchy particles'. Their particles are normally closely looped up but these loops straighten out rather like a telephone cord when the fibre is pulled (figure 24).

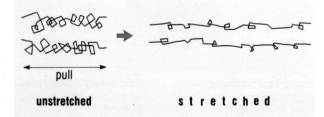

Figure 24 *How a fibre can be stretched*

3 What are polymers?

Fibres have long, thin particles. But how are these long thin particles themselves built up? They are made by joining lots of small units together in a chain. The small units are called **monomers**. The chain is called a **polymer**. The joining-up process is called **polymerisation**. Figure 25 explains the idea. Some polymers have more than one type of monomer unit (figure 26). Most polymers have several thousand monomer units in each chain.

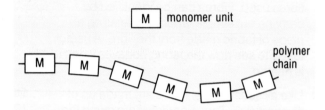

Figure 25 *A polymer made from one type of monomer*

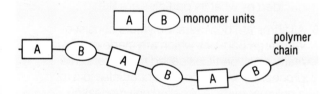

Figure 26 *Part of a polymer made from two types of monomer*

Polymers are named after their monomer. For example, the polymer made from ethene monomer units is called poly(ethene), or polythene.

You can make a model of a polymer chain by joining together paper clips (figure 27). You could call your model 'polypaperclip'!

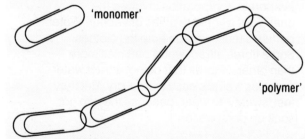

Figure 27 *'Polypaperclip'*

Many natural materials are polymers. Most of your body is made up from the polymers called **proteins**. Carbohydrates such as starch and cellulose are polymers. Polymers can also be made synthetically. Polyester and nylon are synthetic polymers.

4 What holds monomers together in a polymer chain?

Most fibres are **condensation polymers**. Their polymer chains are made by joining monomers together by condensation polymerisation. You have probably done experiments with paper strips to show how condensation polymerisation works. Figure 28 summarises the general idea.

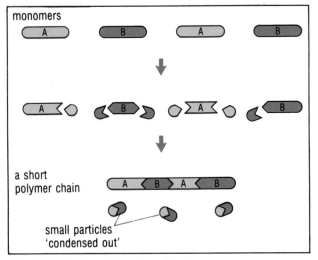

Figure 28 *Condensation polymerisation*

In condensation polymerisation, a small particle is 'condensed out' when the monomer units join together. Often, this small particle is a water particle.

For example, nylon 66 is a condensation polymer. It is made from two monomers called 1,6-diaminohexane and adipic acid. You may have made nylon by reacting these two together in the 'nylon rope trick' (figure 29). Polyesters are also condensation polymers.

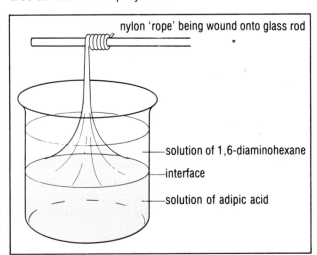

Figure 29 *The 'nylon rope trick'*

Taking it further

So far, we have talked about monomer and polymer particles without saying what these particles are made of.

Monomers and polymers are compounds. They usually contain the elements carbon and hydrogen, and sometimes oxygen and nitrogen as well. Figure 30a shows how atoms of carbon, hydrogen and oxygen are joined together in the monomers that make polyester fibre. Figure 30b shows how these two can join together to form part of a polymer chain.

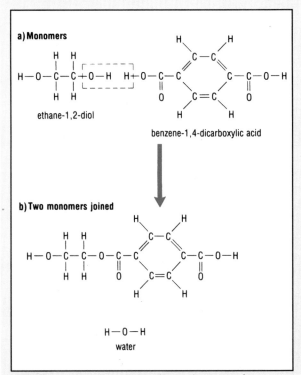

Figure 30 *Monomers join together to form part of a polymer chain.*

Suppose two more monomer units are joined onto the chain. Draw a diagram to show what the new chain would look like.

The properties of a polymer are decided by the nature of the groups of atoms it contains. For example, the groups of atoms in polyester chains tend to attract each other. The atoms on one chain attract those on a neighbouring chain. This holds the chains together quite strongly, which is why polyester makes a strong fibre.

5 Getting clothes clean

Solutions and solubility Figure 31 revises some of the key words which you may have met already to do with solutions.

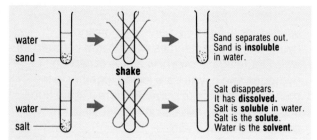

Figure 31 *Words to do with solutions*

Removing grease is the main problem when cleaning clothes. Grease and oil bind dirt onto the clothes. Grease is insoluble in water because grease and water are not attracted to each other. But grease *is* soluble in other solvents, like the solvents used in dry-cleaning.

Even though grease doesn't dissolve in water, you *can* get water to remove grease – by using detergents. Detergents dissolve in both water and grease. They work by loosening grease from clothes so that it can be washed away with water.

6 Wet clothes, dry clothes

After washing clothes we need to get them dry. Have you noticed that some clothes dry more quickly than others? This is because some fibres release the water better than others.

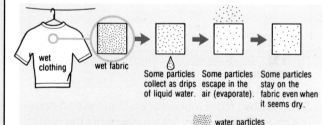

Figure 32 *How clothes dry*

What happens when clothes dry? Look at figure 32. Wet clothes have many water particles on them. Some of these go when liquid water drips off. Some of the water particles escape into the air as vapour – the water evaporates. This happens particularly quickly on a warm breezy day – a good 'drying day'.

Some fibres, like wool and cotton, attract water particles quite strongly. Their fibre particles have a kind of 'water hook' which attracts water particles. Other fibres, like nylon and polyester, attract water particles less. This is because they have fewer 'water hooks' on the fibre particles (see figure 33).

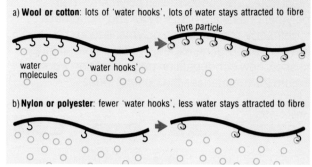

Figure 33 *Fibres with 'water hooks'*

Table 3 compares some of the advantages and disadvantages of fibres which attract water strongly, and fibres which do not.

Table 3 *Comparing the water attraction of different fibres*

	Fibres which attract water strongly	Fibres which do not attract water strongly
Examples	wool, cotton	nylon, polyester
Ease of drying	take longer to dry	dry quickly
Comfortable to wear	comfortable; do not feel 'sweaty' because they absorb perspiration	feel 'sweaty' in hot weather
Crease resistant	crease easily	crease-resistant

Taking it further

Water particles are very simple. Each particle has just two hydrogen atoms and one oxygen atom (figure 34).

Figure 34 *A water particle*

Water particles are attracted to certain other groups of atoms. The O–H and N–H groups attract water strongly. Cotton has particles with lots of O–H groups on, which is why it attracts water. Wool has lots of N–H groups on its particles, which is why wool is water-absorbent.

So 'water hooks' are not really hooks, but attractions between groups of atoms.

Things to do

Things to try out

1 *Using natural dyes*
Try making natural dyes from strongly-coloured plant products like blackcurrants. Crush the coloured materials with warm water – you could use a food mixer. Try dyeing different materials. How effective is your dye? Does it fade? Does it wash out?

2 *Tie-dyeing*
You can buy excellent synthetic dyes in department stores. Try tie-dyeing a white article like a handkerchief or T-shirt. Gather a loop of the material and tie it tightly with thread. Dye the whole article, following the dye manufacturer's instructions. Where the material is tied it will not be dyed, and this can give attractive patterns.

3 *How much water does wool absorb?*
If you have a woollen pullover, try weighing it on a dry day and on a damp day. Is there a difference? If so, why? Do the same for a pullover made from a synthetic fibre like acrylic. How do they compare?

Things to find out

4 How is cotton produced?
5 What is linen? How is it produced?
6 Who made the first artificial fibres, and when?
7 How is soap made? And how does it remove dirt from clothes?

Making decisions

8 *Choosing the right fibre*
Table 4 below gives some properties of four fibres – identified here by the letters W, X, Y and Z.
   a) Which fibre would you use for each of the following jobs? Give the reasons for your choices.
      i) Making a fishing line.
      ii) Making a pullover.
      iii) Making a T-shirt.
   b) You are a manufacturer of socks. You want to make socks out of fibre X because it is warm and absorbs moisture. However, you are worried that socks made of X may wear out quickly. What could you do to get over this problem?
   c) You are a manufacturer of blouses and shirts. You want to make your garments from fibre W because it is comfortable to wear. However, its high moisture absorbency means fibre W creases easily. What could you do to get over this problem?
   d) W, X, Y and Z are actually polyester, wool, cotton and nylon, though not necessarily in that order. Which is which?

Table 4 *Source: Adapted from Take your choice by Norman Reid, University of Glasgow*

Fibre	W	X	Y	Z
Natural or synthetic?	natural	natural	synthetic	synthetic
Strength when dry/g dec^{-1}*	3.2	1.1	4.5	4.1
Stretchiness	low	moderate	moderate	low
Moisture absorbence (%)	9	16	5	0.5
How it feels when worn	comfortable	warm and comfortable	fairly comfortable	fairly comfortable
Durability (how it stands up to wear)	good	moderate	excellent	excellent
Cost	medium	high	medium	medium

*g/dec^{-1} is a measure of force per unit area of cross-section of the fibre

9 Solvents for dry-cleaning

You run a dry-cleaning business. You have the choice of several solvents to use in your dry-cleaning machines, and you need to decide which one to use.

Table 5 below gives important properties of five solvents.

a) Decide which solvent you would use.
b) Explain your reasons for making that choice. (Look back at the section on dry-cleaning on page 49 before you start.)

Solvent	A	B	C	D	E
Ability to dissolve grease	excellent	very good	very good	excellent	excellent
Does it burn?	no	no	no	yes	no
Is its vapour poisonous?	slightly	no	no	no	no
What is its boiling point?	127°C	400°C	130°C	120°C	45°C

Points to discuss

10 Some people think it is wrong to wear clothing made from animals. Many people refuse to wear coats made from animal skins. Some people refuse to wear leather or wool. What do *you* think?

11 Some people don't like wearing synthetic fibres. They say natural fibres feel more comfortable. What do *you* think?

12 Fifty years ago practically all clothes were made from natural fibres like wool, cotton and silk. Today we wear clothes made from many different synthetic fibres as well as natural ones. What kind of clothing materials might we be wearing fifty years from *now*?

13 What matters more – the ways clothes *feel*, or the way they *look*?
a) Suppose someone invented a special spray coating that could be applied to your skin at birth. The coating is warm, waterproof and comfortable to wear. Would there be any more need for clothes?
b) Dyes do nothing to improve the comfort or efficiency of clothes. Yet nearly all clothing is dyed in some way or other. Why do people like their clothes to be dyed? Why do some people like bright colours, and some prefer dull ones? Which do *you* prefer?

Questions to answer

14 Copy out the following paragraph and fill in the missing words. Each word is used only once. The missing words are: cotton, monomers, polyester, polymer, polymerisation, polystyrene, synthetic, wool.
Fibres contain long, thin particles. These long particles are made by joining together lots of small unit particles. The smaller particles are called ___(a)___ and the long chain is called a ___(b)___. The joining up process is called ___(c)___. For example, when lots of styrene particles are joined the product is called ___(d)___. Many fibres are natural. Some come from animals, for example ___(e)___. Some come from plants, for example ___(f)___. Some fibres are not natural, but are made by chemists. They are called ___(g)___ fibres. An example is ___(h)___.

15 Jane was given two small squares of fabric, labelled L and M. She was asked to find out what fibre each was made from. She was told that one of the fibres was synthetic, and one was natural. Jane held each square of fabric in turn close to a Bunsen flame. L melted, but M did not melt.
a) Which fibre was natural and which was synthetic?
Jane now tried burning each fabric in turn. L burned with a yellow sooty flame. There was no strong smell when it burned. M was difficult to burn. When heated strongly it gave a smell of burning hair.
b) Identify L and M.
c) Why is it important for clothing manufacturers to know how fibres behave when heated?

16 a) Explain why water alone is not much use for cleaning greasy clothes.
b) Explain why adding detergent to the water helps it clean.
c) Explain why it is important to move clothes around in the water during washing.
d) Explain why hot water is usually better for washing than cold water.
e) Explain why washing must always be rinsed in clean water before drying.

17 Different clothing fibres have different properties. Four important properties are
A strength   C cost
B water absorbency D stretchiness
Each of these properties is important for a clothing manufacturer to consider when choosing a fibre for a particular use. For each property, give *one* reason why it is important.

FOOD

Introducing food

We all know how important food is in our lives. In Britain we are fortunate that there is no shortage of food. In fact we have a huge variety of foods to choose from.

▲ **Figure 1** *Marvelfood* is a new product. What does the label tell you about it? Is it something you might enjoy? Will it make you fat?

◄ **Figure 2** How long do you think you could last without food and water? Why does your body need food and water?

◄ **Figure 3** How do different foods affect your health? Are these 'health foods' really healthier?

These questions could not be answered fully until chemists were able to analyse food accurately. They could then identify the chemicals in the food you eat and find out what happens to them in your body. The discoveries about the chemical nature of food are among the most important ever for the health and happiness of humankind.

In this chapter you will find out

- ◆ which of the chemicals in food are really necessary,
- ◆ what jobs they do in your body,
- ◆ about the importance of your diet.

Looking at Food

1 Too much, too little, just right?

Whatever age you are and whatever shape or size you are, you need to eat. You need food to give you **energy** for every activity of your life. You even use up energy while you are asleep (figure 4).

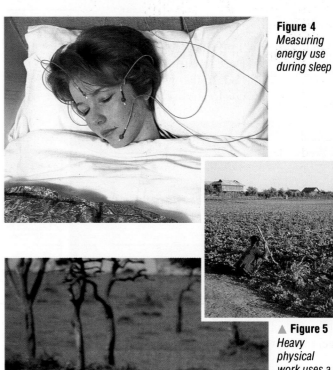

Figure 4 Measuring energy use during sleep

> **1** Think about and then write down what your body uses energy for
> a) when you are asleep,
> b) while you are reading this sentence,
> c) when you are eating,
> d) when you are running.

Food also gives you the chemicals needed to **repair** the cells in your body and to make more of them as you grow.

People who have to hunt for or grow all their food are rarely overweight (figure 5). You do not see fat wild animals, though if food is scarce, you may see very thin ones (figure 6). Not only is it hard for them to get their food, they also use up a lot of energy finding it.

▲ **Figure 5** Heavy physical work uses a lot of energy and keeps you slim.

◀ **Figure 6** The leopard is using a large amount of energy to try and catch a zebra

Millions of people in developing countries have too little to eat. This may be because of famine (not enough food can be grown) or poverty (not enough money to buy what food there is) or other more complicated reasons which you will study later in this course. The problem in the developed countries, such as the UK, is the opposite. Here, there is plenty of food in the shops and people do not use much of their own body energy to bring it home.

Keeping your food intake and your weight just right is not always easy. You need to eat a **balanced** diet which must include carbohydrates, fats, proteins, minerals and vitamins. You also need to keep the correct balance between quantity and quality and this can be difficult. Many people are genuinely addicted to food – especially food containing sugar which is the most fattening and the least useful part of our food intake. The average daily consumption of sugar per person in Great Britain is 125 grams compared with 25 grams in 1880 and 5 grams in 1780 (figure 8).

Figure 7 People who eat too much too often become ill because they are overweight.

Figure 8 Average daily sugar consumption in Great Britain

Of course the best way of keeping your weight just right is never to eat more than your body needs. Judging what your ideal weight should be is quite difficult. Health scientists think that about 30% of the adult population of Great Britain are overweight. What is even more alarming, is that about 20% of school children are also overweight. As **obesity** – being very much overweight – is linked in later life with heart disease, high blood pressure and diabetes (among other illnesses) this is bad news for the nation's health and for individual families.

> 2 If each member of your family eats an average amount of sugar how much sugar would you all eat in a week?
> 3 How much sugar does your family really use in one week?
> 4 Which bought foods that your family eats contain sugar?
> 5 Why do you think your answer to question 3 is different from your answer to question 2?
> 6 Suggest at least one reason why the average daily consumption of sugar has gone up 50 times in the last 200 years.

2 Slimming

If you become overweight there is only one way to get rid of the excess: eat less food than you need. To help you to do this a multi-million pound industry has developed – the Slimming Industry.

There are biscuits to eat to fill you up; there are powders to make into drinks which provide all you need for a meal; there are 'slimmers' soups, crispbreads, cereals, milk, cheese, etc, and there are hypnotists and weightwatching groups.

> Look in the local supermarket or chemists and write down a few examples of slimmers foods and drinks with the 'calorie-content' of each. Make a table with these headings: Name of food, Calories/100g.

None of these slimming aids is any use unless the slimmers eat fewer calories than they need. Food chemists are working on the problem of producing tasty filling food with fewer calories. There is 'low-fat' spread which looks and tastes like margarine but is only half margarine – the other half has no energy value. The biscuits which fill you up before a meal contain methyl cellulose. This is made from the cell walls of plants or from wood pulp; it is like cotton wool. Taken with a drink of water it swells in the stomach and makes you feel less hungry even though it has no energy value because your body cannot digest cellulose.

Chemists are now working on a 'no-fat fat' which is also derived from cellulose. Their aim is to produce a substance which spreads

Figure 9 *The products of the slimming industry*

like fat, tastes like fat but has no calories in it at all! The ideal slimmer's food of the future will be an appetising meal which looks like food, tastes like food, fills you up like food but does absolutely nothing for your energy intake.

3 Artificial sweeteners

Some of the most useful aids to keeping food intake in balance with energy requirements are artificial sweeteners or sugar substitutes. At the moment there are three of these on the market under different trade names. They are all made by chemists and have been extensively trialled to make sure they are safe. None of them contain any 'calories'. The oldest and best-known is **saccharin** which has been used for over fifty years. During the last twenty years a number of other sugar substitutes have come on the market but some have only lasted a year or two before side-effects were reported. **Cyclamate** (sodium cyclamate), for example, gave a natural sweet taste to soft drinks but was withdrawn after a year because, given in very large doses, it was found to cause cancers in rats. **Aspartame** has been available for about ten years but may be withdrawn at any time following reports that it causes dizziness, migraine and other nerve disorders among people using it over a long period. **Acesulphame K** (K because it is the potassium salt of acesulphamic acid) has recently become available.

> If you have a slimmer in your family, look at the special diet foods they eat and list the names and trade names of the artificial sweeteners used.

The chemicals in figure 10 all taste sweet.

But why do some chemicals taste sweet, some taste bitter and others have no taste? The sensation of sweetness is thought to be because there is some sort of reaction between molecules of the chemicals and parts of the surface of the tongue, known as **receptor sites**. The process is very fast and is reversible.

The way the atoms are arranged in the molecules is the key to the process. The symbols ◄ and ⋯⋯ in the formula for aspartame are used to show the three-dimensional form of the molecule. ◄ means the atoms stick up above the page, ⋯⋯ indicates that the atoms lie below the page. Another form of aspartame has these reversed and this compound tastes bitter. The shape of the molecules makes them behave differently.

If you need to lose weight or want to make sure you stay just right, it is probably safer and certainly cheaper to give up extra sugar and sweet foods altogether.

Figure 10 Chemical sweeteners ▼

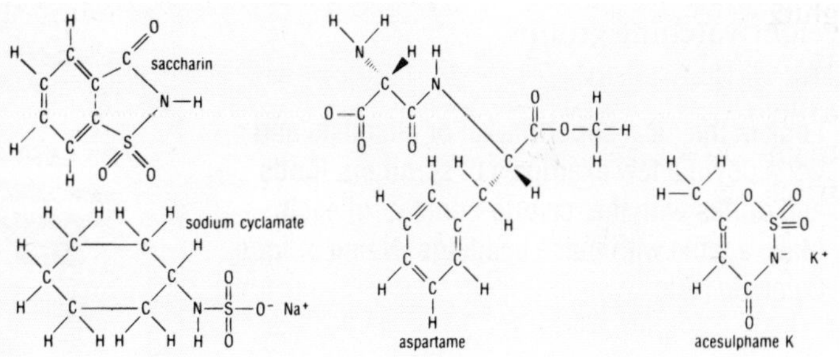

4 Special diets for special people

Diabetic foods

People who suffer from diabetes cannot produce enough insulin. Insulin is a protein hormone which regulates how carbohydrates, such as starch, decompose into smaller molecules including glucose which in turn are used to make energy. Without insulin, the glucose molecules cannot get into the cells of muscle and liver where they are stored for later use. Therefore without insulin, glucose builds up in the blood and is passed out in the urine. The cells do not have enough food and have to feed on the amino acids already in the cells which are normally needed for growth and cell repair. People who have severe diabetes must be given insulin or they will die.

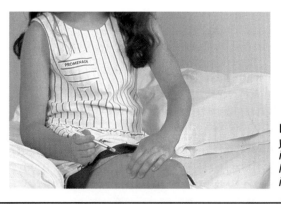

Figure 11 A young diabetic injecting herself with insulin

The insulin given to diabetics has usually been extracted from the pancreas of pigs or cattle. These insulins differ slightly from human insulin. Recently, techniques of genetic engineering have been used to program a bacterium so that it makes human insulin. This insulin is now made on a commercial scale.

One major landmark in the development of this story occurred in 1956 when Fred Sanger (now Sir Frederick) worked out in Cambridge the sequence of amino acids in the insulin molecule. Another occurred in 1969 when Dorothy Hodgkin and her colleagues in Oxford discovered the shape into which the chain of amino acids is folded. These two discoveries have led to a clearer understanding of how insulin works in the body.

Many diabetics do not have to take extra insulin as their bodies do produce some. But all people with diabetes have to be very careful about their diet. They must keep a balance between all the essential foods (carbohydrates, proteins, fats, vitamins and minerals) while strictly limiting their glucose intake.

The sugar used in drinks and for cooking is called **sucrose**. In the body it breaks down to glucose and another sugar called **fructose**.

to the idea that diabetics can use fructose as a safe alternative to glucose. However, some nutritional experts are cautious about the use of fructose and artificial sweeteners may be safer.

Gluten-free foods

Some babies and children are not able to digest one of the proteins in wheat. This protein, called **gluten**, causes stomach upsets and diarrhoea which can be very dangerous for small children. Children with this illness (**coeliac** disease) cannot eat anything made with ordinary (wheat) flour. All baking for them has to be done with rice flour or cornflour.

> 1 Imagine that you had to prepare the food for a coeliac friend for one day. Write down what you would give them to provide a balanced diet.
> 2 Examine a range of items in your store cupboard at home and list those which would be suitable for someone with coeliac disease.

sucrose + water → glucose + fructose

Figure 12
The structures of glucose and fructose

The chemical formulas of glucose and fructose are the same; they are both $C_6H_{12}O_6$. However the way the atoms are joined together is different.

Our bodies recognise the difference. Glucose and fructose are carried into the cells by different mechanisms. This has led

Low-salt diets

We need sodium ions (Na^+) in the body to help move water molecules across cell membranes. This occurs as part of many processes in the body, for example in sweating. Our kidneys regulate the amount of sodium ions in the body. Excess salt is passed out in the urine.

People with heart disease or high blood pressure may be told not to add any extra salt to their food. Table salt (common salt) contains sodium ions (Na^+) and chloride ions (Cl^-).

Because many people find food without salt tasteless, people on low-salt or salt-free diets use 'salt-free salt'. This tastes something like common salt and contains potassium chloride. So it has potassium ions (K^+) and chloride ions (Cl^-). Sodium and potassium ions are transported across cell membranes by different mechanisms and they play different roles in the cell. So potassium ions do not cause the same problems as having too many sodium ions.

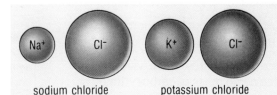

sodium chloride   potassium chloride   **Figure 13**

Baby's first cereals

For the first few months after birth a baby needs nothing but milk (figure 14). Milk provides all the essential chemicals at this age. Once a baby starts to have solid foods it is important to know what is in the foods.

Packets of baby cereals are all carefully labelled to show not only what *is* present (like most packaged food nowadays) but also what is *not* in them (figure 15). These are:

'sugar-free' to try to avoid the baby developing a taste for sweet foods,
'gluten-free' to avoid developing coeliac disease, and
'salt-free' to avoid strain on the baby's growing kidneys.

◀ **Figure 14** *Breast feeding a small baby provides the best nourishment during the first few months of life.*

Figure 15 *Packet of baby cereal showing 'food free' strips* ▶

In brief
Food

1. Your body and all the substances you eat as food are mixtures of chemical compounds. The compounds can contain many different elements but the most common are carbon, hydrogen, oxygen and nitrogen.

2. Foods contain stored chemical energy. They can burn in air to release carbon dioxide, water and heat energy.

3. If you could break down substances into smaller and smaller bits, then eventually you would not be able to break them into anything smaller. These particles are called molecules.

4. When you eat, molecules from your food are broken down by a carefully controlled sequence of reactions involving oxygen. The process is known as respiration. Many chemicals are made and energy is released. Some of the energy is used to build new molecules which act as stores of chemical energy needed to help you to move and grow. Some is released as heat which keeps you warm. The final products of respiration are carbon dioxide and water which you breathe out.

5. A reaction which gives out heat energy is described as **exothermic**.

6. Fats can provide more than twice as much energy as the same mass of carbohydrates or proteins. Unused fats are stored in the body.

7. Simple chemical tests can be carried out to identify proteins, sugars, starches and fats in foods. Tests for pH can show that foods are mildly acidic or neutral or mildly alkaline.

8. The compounds which you must eat to keep healthy are known as **nutrients**. There are seven different types of nutrients.

Essential Nutrients

Carbohydrates (starches and sugars): used to provide energy.

Fats and oils (lipids): stored in the body to provide energy when needed.

Vitamins: protect us from diseases and help the body to make use of other nutrients. Only very small quantities are needed.

Fibre (roughage): not absorbed by the body. Helps to dispose of waste material.

Mineral salts: like vitamins, are required in small quantities. They provide important elements to make more complicated molecules and for other jobs in the body.

Water: our bodies are about 75% water. Almost all the chemical processes which keep us alive and active take place in water in our bodies.

Proteins: used mainly for growth and repair but can provide energy when needed.

Thinking about Chemistry and food

1 What's in food?

In food there are chemicals which you must eat to keep healthy. Other chemicals are added to make food taste and look more interesting and also to make sure the food keeps fresh. You will find out more about these later in your course.

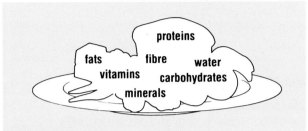

Figure 16 Foods contain seven main groups of chemicals – *nutrients*

Although all the food types in figure 16 are important you need to know about three of them – proteins, carbohydrates and fats – in more detail.

Proteins
The cells and tissues in your body are made of **proteins**. During your life, your body keeps using proteins to replace and repair damaged or dead cells.

All proteins are built from smaller molecules called **amino acids**. These contain carbon, hydrogen, oxygen and nitrogen atoms. Some amino acids, like methionine (figure 17), also contain sulphur atoms.

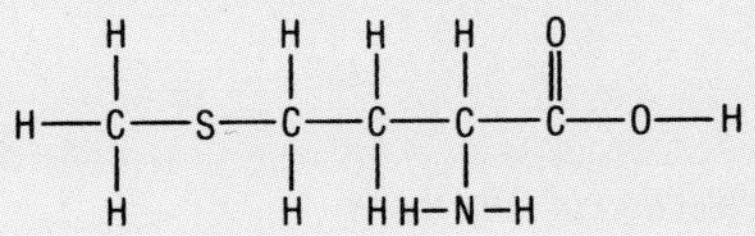

[Chemical structure of methionine]

Figure 17 *Chemical formula of an amino acid – methionine*

Figure 18 shows a small piece of a protein molecule. This piece is made from five amino acid molecules. The amino acids are shown by different shapes. The sequence in which the amino acids are joined is special to each protein.

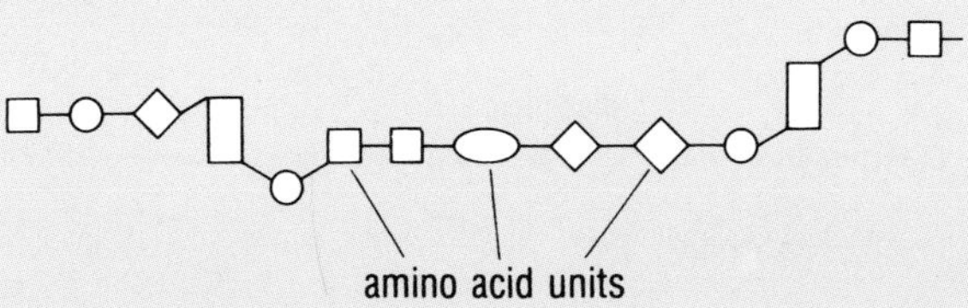

[Diagram of protein chain with amino acid units labelled]

Figure 18 *A piece of a protein chain formed by five different amino acids*

All proteins are made from different sequences of amino acids. With twenty amino acids to choose from, thousands of proteins can be made. They include **haemoglobin**, which is the red colouring-matter in blood, and **keratin** in hair. Haemoglobin is a medium-sized protein molecule, containing about 600 amino acid units. Some proteins contain more than 4000 amino acid units.

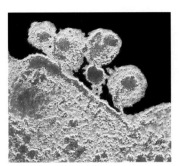

Figure 19 *The AIDS virus (the purple coloured blobs in this electron micrograph) is shown here breaking from an infected cell. The envelope surrounding the virus is made of protein.*

Proteins in the food you eat are broken down to amino acids in your digestive system. The amino acids are carried around your body by your bloodstream to your body cells. Then they join together in various combinations to make new protein chains (figure 20).

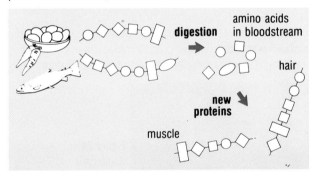

Figure 20

Your body is able to convert one amino acid to another depending on which ones are needed most. But there are eight of the total of 20 amino acid types which it cannot make. If you are to stay healthy, your food must contain proteins which break down to give these eight amino acids. They are called the **essential amino acids**. The richest sources of food proteins which give the essential amino acids are meat, fish, milk, cereals and vegetables, such as peas and beans.

Carbohydrates
Carbohydrates in the form of starches and sugars, are the main source of energy for your body. They are not used for growth or repair. All carbohydrates contain carbon, hydrogen and oxygen.

Figure 21 Bread and wheat ▶

▲ **Figure 22** Rice plant

Figure 23 Rice grains ▶

One important source of starch is cereals – wheat, oats, rice, etc. – from which bread and pasta are made (figures 21 – 23). Another source is vegetables, such as potatoes. These plants make their carbohydrates by the process of **photosynthesis**.

Figure 24 Sugar cane ▼

▲ **Figure 25** A selection of sugars produced from sugar cane

Sugars are the most well-known carbohydrates. To a chemist, **sugars** is the name for a family of substances, all of which have similar properties. The sugar you use to sweeten your food is one of this family (figures 24 and 25). It has the chemical name **sucrose** and formula $C_{12}H_{22}O_{11}$.

Sucrose can be broken down in the body to form two smaller sugars. One is called **glucose** and the other **fructose** (figure 26). They both have the same chemical formula $C_6H_{12}O_6$ and yet they are different substances. Glucose contains a ring of six atoms joined together. Fructose contains a ring of five atoms.

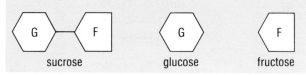

Figure 26

Fats (also known as **lipids**)
Like carbohydrates, fats and oils are important sources of energy for your body. An oil is a fat which is liquid at room temperature.

You eat fats in all sorts of foods (figure 27). There is fat in meat, milk, butter and cheese. These are '**animal fats**'. Many plant seeds and nuts are rich in fats, usually in the form of oil. Sunflower seeds, soya beans, peanuts and olives all provide you with '**vegetable fats**'. To many peoples' surprise, peanuts are about 50% fat.

Figure 27 A selection of fatty foods

Like carbohydrates, fats contain only atoms of carbon, hydrogen and oxygen. But their structures are very different from carbohydrates. Each fat molecule has a 'stem' formed from a chemical called **glycerol**. Attached to the glycerol stem are three molecules of acids known as **fatty acids**. Each of these molecules has a long chain of carbon atoms, with mainly hydrogen atoms attached to it.

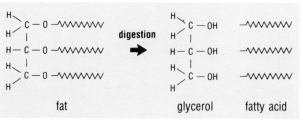

Figure 28

During digestion fats break down into glycerol and fatty acids (figure 28). These move around the body in the bloodstream. They re-combine as new fats which can be stored in the tissues below your skin.

Doctors now encourage people to eat fats which contain fatty acids with some double bonds between the carbon atoms. These fats are known as **polyunsaturates**. They are healthier than saturated fats, which do not have double bonds between the carbon atoms. You will find out more about this later in your course.

2 How do you keep warm?

Have you measured your body temperature with a thermometer? A healthy person should find that their normal temperature is about 37 °C. Except in extremely hot conditions, 37 °C is higher than the air temperature around you. Something inside you is producing heat energy which makes your body warm. Energy is also needed to keep your muscles active and your brain working.

So where do you get the energy from? A clue comes from analysing the air you breathe out (**exhale**). It is not the same as the air your breathe in (**inhale**), see figure 29. The main differences are shown in table 1.

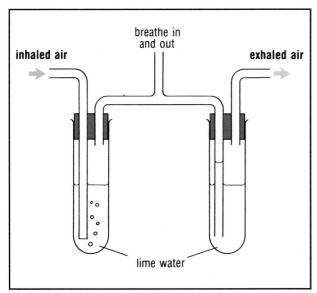

Figure 29

Table 1

	Nitrogen	Oxygen	Carbon dioxide	Water vapour	Other gases
inhaled air/%	78	21	0·03	variable	1
exhaled air/%	78	16	4	always higher	1

Look carefully at table 1 and you will see that something is happening inside your body which:

◆ involves using up some of the oxygen you breathe in,
◆ causes an increase in the carbon dioxide and water vapour in the air you breathe out and raises the temperature.

These changes must be due to the millions of chemical reactions occurring inside you all the time.

In any chemical reaction one set of chemicals (the **reactants**) changes into another set of chemicals (the **products**). During reactions there are always changes in the energy of the chemicals. Sometimes the reaction *takes in* energy from the surroundings. This makes the surroundings colder. Chemists call this type of reaction **endothermic**. More often, chemical reactions *give out* energy, making the surroundings hotter. Chemists call this type of reaction **exothermic**.

There are some endothermic reactions happening inside your body. However most reactions in your body are exothermic as you are usually hotter than your surroundings. These exothermic reactions occur between some of the food chemicals you eat and the oxygen you breath in. The products of the reactions are carbon dioxide, water and energy.

The name given to this whole chemical process is **respiration**. It is a form of **oxidation**. The oxygen you breathe in oxidises the food chemicals.

RESPIRATION:

Food chemicals (Eaten) + Oxygen (Breathed in) → Carbon dioxide + Water + Energy (Breathed out)

Some of the energy from respiration is used to help you grow, move, think and do work. The rest is given out as heat and keeps you warm. So one of the main uses of food is to act as **fuel** for your body. The well known fuels – petrol in car engines and coal on fires – also react with oxygen from the air to produce carbon dioxide, water vapour and energy.

There is a simple demonstration which shows the similarity between the oxidation of food chemicals and other fuels when they burn (see figure 30 on the next page).

If some food or a fuel is burnt, carbon dioxide and water vapour are produced. In this experiment, air is drawn slowly through the apparatus, from left to right.

The presence of water in the products of combustion is shown when the anhydrous copper(II) sulphate turns from white to blue. The lime water turns cloudy to show that carbon dioxide is formed.

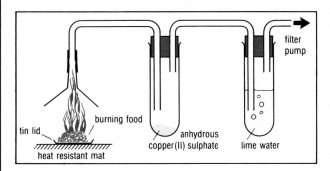

Burning is a very fast set of chemical oxidation reactions between fuel and oxygen. It produces a lot of heat and light very quickly. Fortunately the exothermic reactions which occur during respiration in your body, take place more slowly. If you take heavy exercise, the chemical reactions in respiration speed up. More food is oxidised, which releases energy faster. You feel much hotter. Even so, you will not burst into flames!

3   How can you measure energy?

Energy is normally measured in units called **joules (J)**. It takes 4.2 joules to raise the temperature of 1 gram of water by 1 degree Celsius. In measuring the energy released by burning fuels, it is more convenient to use a larger unit. This is the **kilojoule (kJ)**, which is 1000 joules. A burning match releases about 4 kJ.

You may find another unit which is sometimes used for the measurement of energy from food. This is the **calorie (cal)** or **kilocalorie (kcal)**. The definition of a calorie is that 1 calorie of energy will raise the temperature of 1 gram of water by 1 degree Celsius.

$$1 \text{ cal} = 4.2 \text{ J}$$
$$1 \text{ kcal} = 4.2 \text{ kJ}$$

In modern science only the joule and kilojoule are used. The calorie and kilocalorie appear on food labels and in some books and magazines about nutrition and diet.

Food scientists have found methods for measuring the amounts of energy produced from different foods. This is very difficult to do accurately in a school laboratory. A simple method is shown in figure 31.

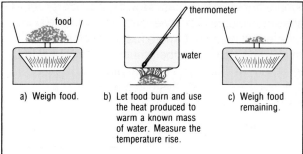

Figure 31

Then calculate as follows:

$$\text{Energy for heating the water from 1 g of food} = \frac{\text{mass of water heated (g)} \times \text{temperature rise (°C)} \times 4.2}{\text{mass of food burned (g)}}$$

For a 1g sample of **sugar** the following results were obtained:

Mass of sugar burned = 1.0 g
Mass of water heated = 200 g
Temperature rise = 20°C

$$\text{Energy released from burning 1g of sugar} = \frac{200 \times 20 \times 4.2}{1000}$$
$$= 16.8 \text{ kJ}$$

When a 1g sample of **margarine** was treated in the same way, the energy released was much greater – 36.6 kJ.

4   What foods give you energy?

All the main food chemicals contain carbon and hydrogen plus other elements. When food chemicals react with oxygen during respiration, the exothermic reaction may be shown as:

Food
$$\begin{matrix} C \\ H \\ O \end{matrix} + O_2 \rightarrow CO_2 + H_2O + \text{Energy}$$

Carbohydrates and fats are the main foods in your body. One gram of carbohydrate releases about 17 kJ during respiration. One gram of fat releases about 37 kJ, more than twice as much. The figure for fat is the same for saturated and unsaturated fats. The low-fat spreads you can buy provide less energy for each gram you eat. This is because water is mixed with the fat!

Your stores of carbohydrate would not last long if they were your only source of energy, for example, even world class marathon runners would run out of energy about halfway through the race if they relied on carbohydrate as a fuel. You should therefore not believe the story that glucose is the main or the most important source of body energy. Nor should you believe that glucose is more readily available than fat. When you are resting, rather more than half your energy is supplied from fat. The proportion changes during activity, but always you are oxidising a mixture of fats and carbohydrates.

5 How can you find out what is in food?

When you are investigating food, many simple tests can be tried in a school laboratory. Only small samples of different foods are needed. Here are some examples of tests you might try.

a) **Examine the food.**
   What is its colour, texture, hardness? Is it like fibre, a jelly or some other form?
b) **Test the food's solubility in water.**
   Does it dissolve quickly or slowly or not at all?
c) **Test the food with a pH indicator.**
   When tested directly or shaken in water the food may show an acidic, alkaline or neutral nature. The pH scale of 1-14 is used for comparing acidity and alkalinity (figure 32). Universal Indicator can show where the pH of a liquid lies on the scale.

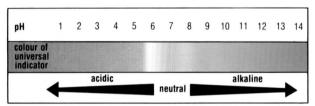

Figure 32 *Universal indicator scale – the pH of a liquid is found by matching the colour on the indicator which has been dipped in the liquid with a reference chart like the one above.*

d) **Heat the food sample gently, then more thoroughly.**
   (i) Does it melt, boil or break down (decompose)? Is any residue left after heating? A black deposit would indicate the presence of carbon in the compounds of the food.
   (ii) Are any gases given off? You should always test for the gas carbon dioxide. If it is present this shows that there are compounds containing carbon in the food.

The test for carbon dioxide is to pass the gas from the heated sample into lime water (calcium hydroxide solution) (figure 33). If the solution goes milky (a white precipitate), carbon dioxide is present.

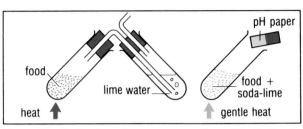

Figure 33 *Testing for carbon dioxide*

Figure 34 *Testing for nitrogen*

e) **Test for water in the food.**
   Water in the food may be present in liquids such as juices. It may also be released when the food is heated gently. Water can be detected by its effect on dry cobalt chloride paper. This changes from pale blue to pink.
f) **Testing for nitrogen in a food.**
   If nitrogen is present, all or most of it will be combined with carbon and hydrogen in proteins. However, when the food sample is heated with soda-lime, the proteins break down (figure 34). Ammonia gas is released. Ammonia contains nitrogen combined with hydrogen (NH_3). Ammonia is an alkaline gas. It turns red litmus paper blue or gives a colour on pH paper which shows a pH well above 7.
g) **Tests for the main food chemicals (nutrients).**
   There are fairly simple tests for starch, glucose, proteins and fats, as shown in figure 35.

Figure 35 *Testing for nutrients*

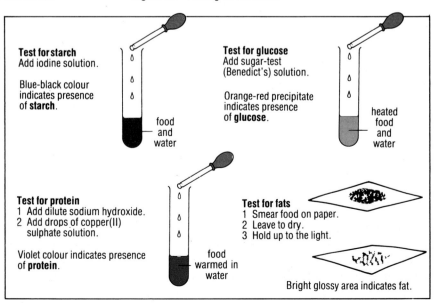

Things to do

Things to try out

1. *How would you detect the presence of carbon dioxide in breathed-out air?*
   Design experiments to investigate if the amount of carbon dioxide breathed out is different before and after eating a meal.

2. *Investigating high-fibre breakfast foods.* One of the properties of fibre (roughage) in your diet is that it absorbs water and swells.
   a) Design experiments to compare the water-absorbing properties of different high-fibre breakfast foods.
   b) Which breakfast food is the most effective absorber of liquids?
   c) Which swells the most?

3. *Indigestion pains are said to be caused by too much acid in the stomach.*
   a) Investigate the anti-acid medicines which you can buy in shops and are used to reduce stomach acidity.
   b) What are the active ingredients in the anti-acids?
   c) Try adding crushed anti-acid tablets to an acidic substance such as lemon juice or vinegar. Does the acidity disappear?
   d) Design experiments to find out which anti-acid is the most effective.

Things to find out

4. a) What is the modern chemist's name for 'bicarbonate of soda'?
   b) Why can it be used as a raising agent in the baking of bread or cakes? What are the chemical reactions involved?

5. How is sugar extracted from sugar beet on a large scale? Are you able to obtain sugar crystals from a sugar beet in your laboratory?

Points to discuss

6. What sort of foods should be packed in survival kits for storing in aeroplanes or lifeboats? The people who use them may have to live without any means of cooking for about a week, in very difficult conditions, such as in an open boat or in the desert.

7. How would you convince people in your school that there are good reasons for eating food which is high in fibre and low in fats?

8. Someone says 'Food is just a lot of chemicals we happen to eat. It would be much easier and healthier if we could get all our food as a few pills every meal.' What do you think about this idea?

Questions to answer

9. The labels on five jars containing white powders in a food cupboard have fallen off and become mixed up. The labels say

   protein extract   starch   glucose   salt   baking soda

   How would you find out which substance is in each jar?

10. A TV advertisement for a drink claims that 'it gives you energy'. A bottle containing 1 kg of the drink costs 99 pence. On the label of the bottle you read that the liquid contains 'glucose 20%, citric acid 0.5%, lactic acid 0.1%, vitamin C 0.2% and carbonated water'.
    a) What does 'carbonated water' mean? How would you test the drink to prove your answer?
    b) What is the chemical source of energy in the drink?
    c) What mass of this chemical is there in the bottle?
    d) How much would the same mass of this chemical cost if you bought it in pure form from a shop?
    e) What is your opinion about the value of this drink as a source of energy?

11. The energy we use in various activities has been measured.

Activity	Energy used/ kilojoules per minute	Activity	Energy used/ kilojoules per minute
sleeping	4	cycling	25
sitting	6	dancing	25
standing	7	running	35
walking slowly	13	football	35
walking quickly	21	swimming	35

You will also have seen tables of data about the energy provided by different foods.
a) From the time you get out of bed until lunchtime, calculate approximately how much energy you have used.
b) Calculate how much energy you gained from your breakfast.
c) Decide if you are eating too much or too little of the right type of food at breakfast.
d) One cream meringue provides about 850 kJ. To use up just this energy, for how long would you have to
   (i) sit still,   (ii) walk quickly,
   (iii) take vigorous exercise?

INDEX

If more than one page number is given, you should look up the **bold** one first.

Acesulphame K, 50
Acid rain, **28-9**, 32
Acidic oxides, **32**
Acids, **32**
Adipic acid, 43
Air, burning in – see Burning
   composition of, **30**, 55
   exhaled air, 55
   pollution of, **28-9**, 32
Alcohol, drinking laws, **14-15**
   effect on health, 2, **14-15**
   in drinks, **15**, 19
   making drinks, 15, **19**
   oxidation of, 15
Alkali, **32**
Alloys, 1, 7, 8, **10**
Aluminium, corrosion of, **5**
Amino acids, essential, **53**
   in proteins, 51, **53**
Ammonia, test for, 57
Aramid fibre, 38-9
Aspartame, 50

Batteries, car, 3
Brass, **9**
Bronze, 1
Bunsen burner, 25-6
   flame, 25-6
Burning, **24-7**, 30
   acetylene, 30
   methane, 25
   petrol, 30
   to form carbon dioxide, 25, 30; 31
   to form carbon monoxide, 31

Calcium, reaction with water, 9
Calcium oxide, 28
   use in power stations, 28
   use in water purification, 18
Calcium sulphate, produced in power stations, 28
   use in making plaster, 28
Calorie, 56
Carbohydrates – see also Fructose, Glucose, Starch
   as polymers, 42
   making alcoholic drinks, 19
Carbon, in steel, 10
Carbon dioxide, added to drinks, 17
   formed during fermentation, 20
   formed in burning, 25, 29, 32
   properties of, 17
   test for, 19, 56, **57**
   use in fire fighting, 27
Carbon monoxide, as a poison, 31
   formed in burning, **26**, 31
   from car exhausts, 31
Cellulose, in cotton, 40, 49
   in paper, 40
   in wood, 49
   to make foods, 49
Chemical energy – see Energy
Chemical reaction, 29
Chlorine, properties of, **17**
   uses of, 17, 19
Chlorophyll, 16
Cigarettes, producing carbon monoxide, 31
   toxicity of nicotine, 2
Cleaning, clothes, 37, 44
Clothing, **35-46**
   protective, **38-9**
Coal, as a fuel, 25, 28
   coke, 29
   combustion, 29, 30
   compound, 7, 9
   see also Burning
Condensation, 17, 18, 37
   polymerisation, 43
Copper, in alloys, 9, 10
Corrosion, **4-5**, 7, 29
   prevention of, **4-5**, 7
Cotton, 39, 44
Crystals, in metals, 6
Cyclamate, 50

Detergents, 41, 44
Diabetes, 50-1
1, 6-diaminohexane, 43
Distillation, 17, 19-20, 37
Drinks, **13-22**
Dry cleaning, 37

Earth's crust, composition of, 11
Electricity, as a fuel, 24-5
Elements, 7, 9-10
Endothermic reactions, 54
Energy, conversion of, **30**
   from food, 23, 48-9
   from fuels, 24
Ethene, use in making polythene, 42
Evaporation, 17, 18, **41**, 44
Exhaust, from cars, 3, 31
Exothermic reactions, 29, 52, 54

FGD (flue gas desulphurisation), **28**
Fabric, 35, 40
Fats, in food, 54-5
Fatty acids, 54
Fermentation, 16, 17, **19-20**
Fibre, in food, 52
Fibres, cleaning of, 37, 44
   dyeing of, 36
   fire resistant, 38-9
   properties of, 42, 44
   treatment of, 41
   types of, 41
Fire, fighting, 27, 38
   protective clothing, 38-9
   triangle, 27
Flue gas desulphurisation (FGD), **28**
Fluoridation, of water, 18
Food, 30, **47-58**
   as a source of chemical energy, 30, **48-51**, 53-7
   for babies, 51-2
   for diabetics, **50-1**
   tests for, 57
Fructose, 51, 54
Fuels, as a source of chemical energy, 30
   burning of – see Burning
   comparison of costs, 25
   comparison of properties, 29
   different fuels, **24-5**

Gas – see Natural gas
   cookers, 25-6
Glucose, 50-1
Gluten, 51-2
Glycerol, 54
Gold, 4
Gypsum – see Calcium sulphate

Haemoglobin, as a protein, 53
   in blood, 31, 53
   reaction with carbon monoxide, 31
   reaction with oxygen, 31
Halon, use in fire fighting, 27
Hydrogen, formation from water, **8-9**
Hydroxides of metals, 8-9

Insoluble substances, 17
Iron, alloys, 10
   corrosion of, **4-5**, 7
   protection by tin, 4
   protection by zinc, 4
   protection of magnesium, 4

Joule, 56

Kilocalorie, 56
Kilojoule, 56

LD 50, 2
Lead, as a poison, 2-3
   in alloys, **10**
   use in petrol, 3
Lethal dose (LD), 2
Lime – see Calcium oxide

Magnesium, use in corrosion, 4
Maltose, 19
Mauve, 36
Metals, **1-12**, 32
   construction from, 7
   corrosion of, **4-5**, 7
   hardness of, **6-7**
   physical properties of, 6, 7

poisonous nature of, **2-3**
reaction with oxygen, 7
toxicity of, **2**

Methane, use as a fuel, 25
Molecule, **52**
Monomer, 41

Nappies, 39-40
Natural gas
  – *see also* Methane, 25
Nicotine, toxicity of, 2
Nitric acid, in acid rain, 32
Nitrogen, in air, 30, 55
  test in food, 57
Non-metals, 7, **32**
  reaction with oxygen, 32
Nylon, **42**, **43**, 44
  rope trick, **43**

Oxidation, of alcohol, 15
  of food, **55-6**
  of fuels, **29**
  see also Respiration, Rusting
Oxides, of metals, 7, **32**
  of non-metals, **32**
Oxy-acetylene torches, 30
Oxygen in air, 30, 55
  reaction with acetylene, 30
  reaction with alcohol, 15
  reaction with calcium sulphite, 28
  reaction with iron and water, 4, 7
  reaction with metals, 7
  reaction with methane, 25
  reaction with natural gas, 25
  reaction with non-metals, 32
  reaction with petrol, 30, 31

pH scale, **57**
Paints, 3
Particulate theory, 17, **20**
Periodic Table, **10**, **32**
Petrol, burning of, 30-1
  formation of carbon dioxide, 30

Poisons, alcohols, **14-15**
  metals, **2**
  nicotine, 2
Pollution, by acid rain, **28-9**, 32
  by lead, **3**
Polyesters, **43**, **44**
Poly(ethene) – *see* Polythene
Polymers, **41**, **42-3**
Polymerisation, **41**, **42-3**
Polythene, **42**
Polyunsaturated fats, 55
Potassium chloride, in diet, 51
Power stations, clean
  atmosphere, **28**, **32**
  energy conversion, **30**
Proteins, in diet, **53**
  made up of amino acids, 51, **53**

Quicklime – *see* Calcium oxide

Reactivity series, **5**, **7**, **8-9**
Respiration, 52, **55-6**
Rusting, **4**, **7**, **8**
  protection from – *see* Iron

Saccharin, 50
Salt – *see* Sodium chloride
Smoking, forming carbon
  monoxide, 31
  nicotine, 2
Sodium chloride, in diet, 51-2
Sodium cyclamate, 50
Solder, **10**
Solubility, **44**
Solute, 17, **44**
Solution, **44**
Solvent, 17, 37, **44**
Soot, 31
Starch, 19, **53-4**
Steel, as an alloy, 10
  cutting of, 30
Sucrose, 19
  consumption in UK, 48
  reaction to form glucose, 51, 54

Sugar – *see* Sucrose
Sulphur, in coal, 28, 29
Sulphur dioxide, as a pollutant, **28-9**, **32**
  formation of, **28-9**, **32**
Sulphuric acid, in acid rain, 32
Sulphurous acid, in acid rain, 32
Sweeteners, artificial, 50
Symbols, of elements, **7**, **8**, **10**

Tea, consumption of, 16
  production of, 16
Terylene, 43
Tetrachloroethene, 37
Tetraethyl lead, 2
Thread, 41
Tin, alloys of, **9**, **10**
  use in corrosion, 4

Vitamins, 52

Warmth, **23-34**
  for houses, **24-5**
Water, **13-22**. 32, 40, 57
  absorption by cotton and
paper, 52
  addition of chlorine to, 17
  addition of fluoride to, **18**
  addition of lime to, **18**
  boiling, 20
  fluoridation of, **18**
  formation of hydrogen from, **9**
  purification of, 13, 17, **18**
  reaction with metals, **9**
  reaction with non-metal oxides, **32**
  shortage of. 1, 17
  test for, 56, **57**
Water cycle, **18**
Wood, as a fuel, 24
Wool, 39, 42, 44

Yeast, 19

Zinc, use in corrosion, 4